FOOD SCIENCE AND TECHNOLOGY

ONION CONSUMPTION AND HEALTH

FOOD SCIENCE AND TECHNOLOGY

Additional books in this series can be found on Nova's website under the Series tab.

Additional e-books in this series can be found on Nova's website under the e-book tab.

NUTRITION AND DIET RESEARCH PROGRESS

Additional books in this series can be found on Nova's website under the Series tab.

Additional e-books in this series can be found on Nova's website under the e-book tab.

FOOD SCIENCE AND TECHNOLOGY

ONION CONSUMPTION AND HEALTH

CANDELARIO B. AGUIRRE
AND
LETICIA M. JARAMILLO
EDITORS

Nova Science Publishers, Inc.
New York

For permission to use material from this book please contact us:
Telephone 631-231-7269; Fax 631-231-8175
Web Site: http://www.novapublishers.com

NOTICE TO THE READER

Additional color graphics may be available in the e-book version of this book.

LIBRARY OF CONGRESS CATALOGING-IN-PUBLICATION DATA

Onion consumption and health / editors, Candelario B. Aguirre and Leticia M. Jaramillo.
p. ; cm.
Includes bibliographical references and index.
ISBN 978-1-62100-836-1 (hardcover)
I. Aguirre, Candelario B. II. Jaramillo, Leticia M.
[DNLM: 1. Onions. 2. Flavonoids. 3. Food Analysis. 4. Phytotherapy. WB 430]

635.25--dc23

2011039221

Published by Nova Science Publishers, Inc. † New York

CONTENTS

PREFACE

Flavonoids and organosulfur compounds are the two major classes of secondary metabolites found in onions that are believed to play a role as health-promoting, disease-preventing dietary supplements, including antioxidant activity. In this book, the authors present topical research in the study of onion consumption and health effects. Topics include an overview on the bioactivity of onions; an ultrasound-assisted flavonoid extraction from onions; and the physiological attributes of GST, Gly I, Gly II and allinase in onions bulbs and cultured cells.

Chapter 1 - Onion (*Allium cepa* L.) is an important vegetable traditionally used as a food ingredient in the Mediterranean diet that has a high production, domestic, and foreign trade worldwide. It is consumed raw, cooked or processed into different onion products in the daily diet. Onion added into different foods makes these products rich in bioactive compounds with potential beneficial health effects. Among them, its effect on cardiovascular disease, including hypocholesterolemic, hypolipidemic, anti-hypertensive, antithrombotic, and hypoglycemic activities, is one of the most extensively studied benefits. Onion consumption has also been reported to have antiproliferative effects in many cancer cell lines, to be involved in the bone metabolism and in the behaviour as a possible antidepressant agent, and to stimulate the growth of specific microorganisms in the colon (*Bifidobacteria* and *Lactobacilli*) with a general positive health effect. Moreover, traditionally, in the folk medicine, it has been described the use of onion as an antimicrobial, antioxidant, anti-inflamatory and asthma-protective agent. Evidence from several investigations suggests that these biological and medical functions are mainly due to the high content in organo-sulphur compounds content of onion. Along with organo-sulphur compounds, organo-selenium compounds, flavonols (quercetin and its glucosides) and dietary fibre (fructans and fructooligosaccharides (FOS)) have been also related to the onion biological properties. Moreover, recently, it has been demonstrated that additional onion constituents such as saponins and peptides have potentially beneficial health effects, including antifungal, antitumor, antispasmodic and cholesterol-lowering activities and capacity to inhibit *in vitro* the development and activity of osteoclasts.As with every biologically active substance, with onion and its derivatives it is necessary to consider certain precautions to minimize the risk of adverse side effects. However, the usefulness of onion as therapeutic agent seems to be very safe, since all its possible adverse effects, such as gastrointestinal upsets and dermatological problems appear with an excessive and prolonged consumption.

Chapter 2 - Onions are valued globally for their medicinal attributes. Flavonoids and organosulfur compounds are the two major classes of secondary metabolites found in onions that are believed to play a role as health-promoting, disease-preventing dietary supplements, including antioxidant activity. Onion (*Allium cepa* L.) bulbs contain high levels of glutathione *S*-transferases (GSTs), glyoxalase I (Gly I), glyoxalase II (Gly II) and alliinase activities that are associated with quality, taste and aroma of foodstuffs. GSTs are involved in the intracellular detoxification of xenobiotics and endogenous toxins whereas Gly I and Gly II detoxify methylglyoxal (MG), a toxic compound produced under normal physiological conditions. Alliinase, on the other hand, is involved in the metabolism of sulfur compounds derived from glutathione (GSH). The interaction of onion bulb GSTs, separated by DEAE-cellulose chromatography, with sulfur compounds has been investigated. Two non-physiological compounds, *S*-hexyl GSHand *S*-butyl GSH, strongly inhibited the 1-chloro-2,4-dinitrobenezene (CDNB)-conjugating activity of GSTa, GSTb and GSTe. However, physiological sulfur compounds, *S*-methyl GSH, *S*-propyl GSH, *S*-lactoyl GSH and *S*-ethyl-L-cysteine sulfoxide, have small or almost no inhibitory effects. Therefore, onion sulfur compounds might have the least possibility to be the substantial physiological counterparts of onion GSTs. On the other hand, the activities of GSTc, GSTd and *Ac*GSTF1 are strongly inhibited by flavonoids, quercetin, luteolin, apigenin and kaempferol. The ethylacetate extract of onion bulbs contains quercetin-4′-glucoside as a major inhibitory substance, whose strong inhibitory effects on GSTc, GSTd and *Ac*GSTF1, as well as its high concentration in onion bulbs, indicate that it is a physiological counterpart of dominant GSTs in onion bulbs. GSTe was slightly inhibited by quercetin, but the activity showed a significant seasonal variation between onion bulbs harvested in autumn and spring, suggesting that this GST isoform might play a role in abiotic stress tolerance through an unknown function. The authors purified and characterized onion GST and Gly I proteins and studied their expression pattern under various abiotic stresses. Three types (short-, medium- and long-) of onion Gly I cDNA were also isolated, cloned and sequenced. However, only short-type Gly I cDNA showed enzymatic activity. The two other types of onion Gly I had no enzymatic activity, although their presence indicates that they might have other physiological functions. The evolutionary relationship showed that medium- and long-type Gly I evolved from the gene duplication of short-type Gly I sequences. In this chapter, they will provide an overview of the physiologicalattributesof GST, Gly I, Gly II and allinase in onion bulbs and cultured cells. The possible relationship among these enzymes as well as their physiological inhibitors and in relation to stress tolerance will also be discussed.

Chapter 3 - Since ancient times, onions, *Allium cepa* L., one of the most widely distributed vegetables, have been used as common foods and for the treatment of many diseases. Onion is a food widely used in gastronomy and, also, for medicinal applications. It is one of the most cultivated and consumed vegetables on the planet (about 66 million tons in 2008) and one of the main ingredients of the Mediterranean diet. This plant is a rich source of several phytonutrients recognized as important elements of the Mediterranean diet and it is known as a good source of bioactive compounds, such as sulphur-containing compounds and flavonoids. Carbohydrates in onion constitute about 80% of the dry matter, and the major non-structural carbohydrates of the onion bulb are fructo-oligosaccharides, well known as fructan, followed by glucose, fructose and sucrose.. Various protein sources can be found in intact onion in low concentrations, including lectins (the most abundant proteins in onion), prostaglandins, ... Also, can be found vitamins (B1, B2, B6, C, E, biotin and nicotinic acid), fatty acids, glycolipids, phospholipids and essential amino acids. But, the most recent attention has focused on the flavonoids that are also responsible for a great part of the health benefits. It is known that onions are a rich source of dietary flavonoids with average content ranging from 270 to 1917 mg of flavonoids per Kg of fresh weight, depending on the onion variety. They are mainly represented by the flavonols quercetin and kaempferol and other flavonoids such as isorhamnetin, myricetin and their conjugates (as glycosylated forms). Anthocyanins, namely peonidin 3-glucoside, cyanidin 3-glucoside and cyanidin 3-arabinoside and their malonylated derivatives, cyanidin 3-laminariobioside and delphinidin and petunidin derivatives, are located in the skin of red onions and in the outer fleshy layer. Previous studies indicate that flavonoids have beneficial health effects due to their antioxidant, anti-inflammatory, cardioprotective, and anticancer effects, which are of particular interest in the prevention of chronic diseases, including cardiovascular disease and some cancers. Onion extract has been recently reported to be effective in reducing risk of cardiovascular disease because of its hypocholesterolemic, hypolipidemic, anti-hypertensive, antithrombotic and anti-hyperhomocysteinemic effects, and to possess many other biological activities including antimutagenic, antiasthmatic, immunomodulatory and prebiotic activities. For this reason can be used as alternative to additive because nowadays consumers are demanding minimally processed foods without synthetic preservatives, due mainly to concerns about possible adverse health effects. It should be noted that flavonoids of the onion are more stable than the other components. Traditionally, it was suggested that the sulfur compounds are those that are good for health because they are responsible for the characteristic flavour, aroma and tear promoting effects of the plant. These substances, highly volatile and unstable, are released when the onion is damaged or cut. The most important organosulfur compounds are the cysteine sulfoxides and the gamma-glutamylcysteines. Onions and onion by-products have been proposed as food ingredients which can be added to food preparations to enhance their preservation during their shelf-life. Onion is widely accepted as a good alternative as a food preservative, especially in fresh or minimally processed foods. The organosulfur compounds have been shown prevent formation of reactive oxygen species and reduction of cholesterol levels. Epidemiological studies have shown an inverse relationship between consumption of vegetables and the risk of cardiovascular disease. The beneficial effect of flavonoids in protection against cardiovascular disease was demonstrated. By conclusion, the onions are very suitable as an ingredient in food preparation or for direct consumption. There has been increased consumption of this vegetable because of a greater public interest in natural remedies and non-conventional therapies for the treatment of certain diseases. Many consumers believe that these therapies have fewer side effects than pharmaceutical products. In spite of numerous studies about the beneficial effects of onions on human health, certain aspects still need to be investigated. Onion could be added like functional ingredients into ”fast foods”, which are widely consumed in current society, to provide them with antioxidant compounds, such as quercetin, prebiotics, or mineral nutrients, such as selenium, to prevent nutritional deficiencies.

Chapter 4 - Nowadays, there is a constant and increasing social concern about the food we, as humans, daily eat and particularly about vegetables.

Moreover, there is a worldwide great demand of minimally processed food or food ingredients with functional properties as a consequence of the changing healthy and environmental sustainable living that is becoming to spanning almost the totality of the world's cultures. The Mediterranean diet which traditionally includes generous amounts of fruits and vegetables has been associated with health benefits such as lower risks for cardiovascular disease (CVD) and cancer. There is epidemiologic evidence that frequent consumption of *Allium* vegetables such as garlic, scallions, onions, chives, and leeks is protective against cancer. Therefore, *Allium* vegetables should be considered to be an important component of a healthy, cancer and CVD resistant diet. The authors consider that it is needed to study in depth not only the technological but also the biological and nutritional properties of vegetables products in order to better understand their role in humanity's daily diet and lifestyle. In this chapter onion has been chosen as a model of an *Allium* vegetable due to its good technological and nutritional properties reported and to its presence in the daily diet worldwide. Onion is a good source of bioactive compounds such as flavonoids, organosulfur compounds (OSCs), fructans and fructooligosaccharides (FOS). Onion has been described to have several health benefits related to its antioxidant, anticarcinogenic, hypolipidemic, hypoglycaemic, or antiaggreagatory effects.

Chapter 5 - Onion (*Allium* cepa L.) is a food ingredient widely used in worldwide gastronomy, making a significant nutritional contribution to the human diet. Moreover, since ancient times, onions have been used as medicinal agents. In recent years, extensive research has focused on the beneficial and medicinal properties of onions and it has been reported that onions are effective in cardiovascular disease, due to their hypocholesterolemic, hypolipidemic, anti-hypertensive, anti-diabetic, antithrombotic and anti-hyperhomocysteinemia effects, and have many other biological activities such as antimicrobial, antioxidant, anticarcinogenic, antimutagenic, antiasthmatic, immunomodulatory and prebiotic activities. These properties are due to the presence in onions of several groups of compounds such as flavonoids, alk(en)yl cysteine sulphoxides, fructooligosaccharides and dietary fiber among others. Thus, onions and their derived products, such as onion skins, two outer fleshy scales and roots, constitute an interesting source of dietary fiber, phytochemicals and natural antioxidants, so that their application in food, which would increase health promoting properties of foods, is a promising field. However, the composition varies throughout the bulb and also is dependent of the cultivar. Therefore, in the present article, healthy properties of products derived from onions have been studied and reviewed in relation to their composition, in order to assess their possible use as functional ingredients rich in specific compounds. Such information may be useful to food technologists for the appropriate exploitation of onion products as source of a specific functional compound.

Chapter 6 - Due to the use of ambient temperature (25°C) and water having pH 7 during extraction step, ultrasounds have been found to be an interesting alternative to conventional methods for antoxidants extraction. Extracts obtained in presence of ultrasound waves have shown clearly their impact in increase of total dry extract yield along with increase extraction of flavonol contents. Another advantage of using ultrasounds is the rapidity of the extraction. This extraction step require several hours in conventional processes, but it only required few minutes in ultrasounds assisted systems which induced better recovery of flavonol contents.

In: Onion Consumption and Health
Editors: C. B. Aguirre et al.
ISBN: 978-1-62100-836-1

Chapter 1

AN OVERVIEW ON BIOACTIVITY OF ONION

Marta Corzo-Martínez and Mar Villamiel*
Instituto de Investigación en Ciencias de la Alimentación,
CIAL (CSIC-UAM) C/ Nicolás Cabrera.
Campus de la Universidad Autónoma de Madrid, Spain

ABSTRACT

Onion (*Allium cepa* L.) is an important vegetable traditionally used as a food ingredient in the Mediterranean diet that has a high production, domestic, and foreign trade worldwide. It is consumed raw, cooked or processed into different onion products in the daily diet. Onion added into different foods makes these products rich in bioactive compounds with potential beneficial health effects. Among them, its effect on cardiovascular disease, including hypocholesterolemic, hypolipidemic, anti-hypertensive, antithrombotic, and hypoglycemic activities, is one of the most extensively studied benefits. Onion consumption has also been reported to have antiproliferative effects in many cancer cell lines, to be involved in the bone metabolism and in the behaviour as a possible antidepressant agent, and to stimulate the growth of specific microorganisms in the colon (*Bifidobacteria* and *Lactobacilli*) with a general positive health effect. Moreover, traditionally, in the folk medicine, it has been described the use of onion as an antimicrobial, antioxidant, anti-inflamatory and asthma-protective agent.

---

* Author to who correspondence should be addressed: Tel +34 910017951; Fax +34 910017905. E-mail: m.villamiel@csic.es

Evidence from several investigations suggests that these biological and medical functions are mainly due to the high content in organo-sulphur compounds content of onion. Along with organo-sulphur compounds, organo-selenium compounds, flavonols (quercetin and its glucosides) and dietary fibre (fructans and fructooligosaccharides (FOS)) have been also related to the onion biological properties. Moreover, recently, it has been demonstrated that additional onion constituents such as saponins and peptides have potentially beneficial health effects, including antifungal, antitumor, antispasmodic and cholesterol-lowering activities and capacity to inhibit *in vitro* the development and activity of osteoclasts.

As with every biologically active substance, with onion and its derivatives it is necessary to consider certain precautions to minimize the risk of adverse side effects. However, the usefulness of onion as therapeutic agent seems to be very safe, since all its possible adverse effects, such as gastrointestinal upsets and dermatological problems appear with an excessive and prolonged consumption.

Keywords: Onion, health, beneficial effects

INTRODUCTION

Onion (*Allium cepa*) is original from central Asia and is one of the oldest cultivated plants since it has been in cultivation for more than 4000 years. Onion and other species of the genus *Allium* have been traditionally used worldwide for various purposes such as food preparation and seasoning agents. In particular, the importance of onion lies in the flavour that it imparts to other foods due to its composition. The main constituents of onion are shown in Table 1.

It is usually consumed as fresh, however since losses of fresh onion in storage have been reported to be about 20-30% (Chadha and Sidhus, 1990), processed products are the most practical solution. Thus, the international market of onion is increasingly focused mainly on dehydrated products such as flakes, rings, granules, kibbles, powder, etc. and frozen or canned onions, or onion in vinegar, in brine or as essential oil, its commercial products being less abundant than those of garlic.

Dehydrated products have great commercial value not only by their culinary use but also their by medicinal properties as nutraceutical since they contains higher concentrations of beneficial compounds than the fresh forms (Lanzotti, 2006).

Table 1. General composition of onion

Energy	23-38 Kcal 100 g^{-1} fresh weight
Protein	0.9-1.6 %
Fat	Trace-0.2%
Carbohydrates	5.2-9.0%
Ash	0.6%
Vitamins:	mg 100 g^{-1} fresh weight
Vitamin C	10.0 mg
Vitamin D	0.3 mg
Riboflavin	0.05 mg
Biotin	0.9 μg
Nicotinic acid	0.2 mg
Folic acid	16.0 μg
Pantothenic acid	0.14 μg
Elements:	mg 100 g^{-1} fresh weight
Ca	190-540
P	200-430
K	80-110
Na	31-50
Mg	81-150
Al	0.5-1
Ba	0.1-1
Fe	1.8-2.6
Sr	0.8-7
B	0.6-1
Cu	0.05-0.64
Zn	1.5-2.8
Mn	0.5-1.0
S	50-51

In China, onion tea has long been recommended for several pathologies such as fever, headache, cholera and dysentery. Evidence from several investigations suggests that the biological and medical functions of onions are mainly due to their high content in organo-sulphur compounds (Augusti and Mathew, 1974; Wargovich *et al.*, 1988). The primary sulphur-containing constituents in this vegetable are the S-alk(en)yl-L-cysteine sulphoxides

(ACSOs), such as alliin, and γ-glutamylcysteines, which, besides to serve as important storage peptides, are biosynthetic intermediates for corresponding ACSOs from which, and by different metabolic pathways in each vegetable, volatile, such as allicin, and lipid-soluble sulphur compounds, such as diallyl sulphide (DAS), diallyl disulphide (DADS) and others, are originated (Lancaster and Shaw, 1989). These compounds provide to onion its characteristic odour and flavour, as well as most of its biological properties (Lanzotti, 2006) (Figure 1).

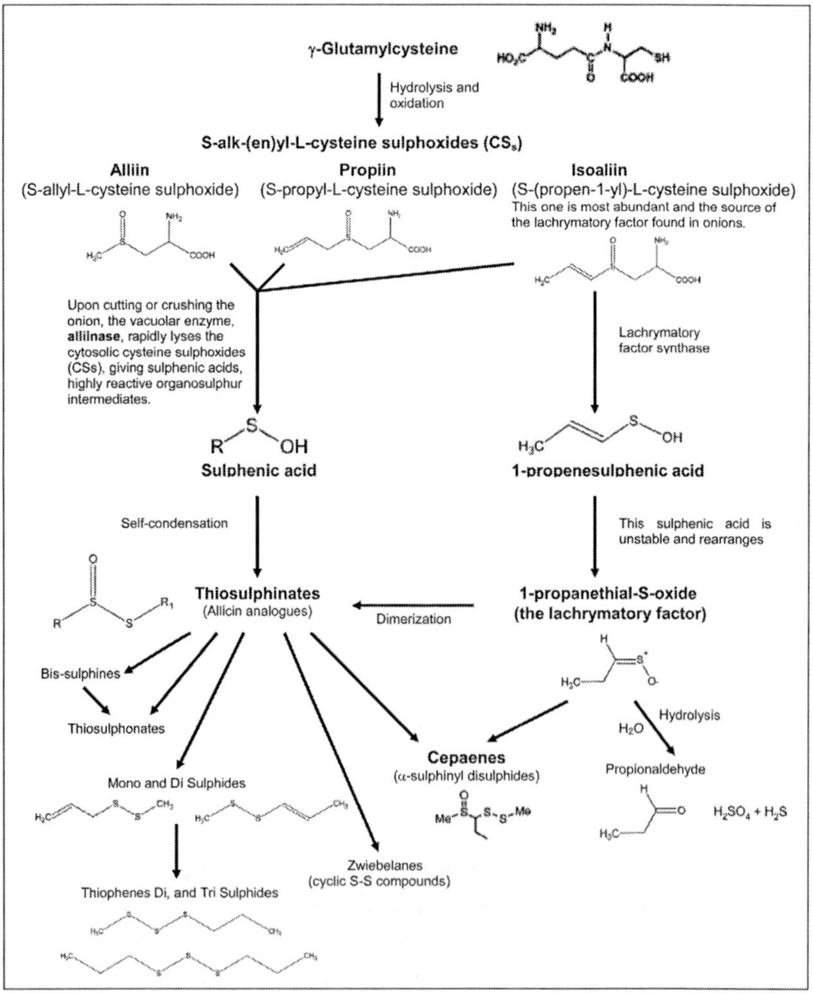

Figure 1. Formation of organo-sulphur compounds during metabolic pathways in processed onion (Taken from Corzo-Martínez et al. 2007).

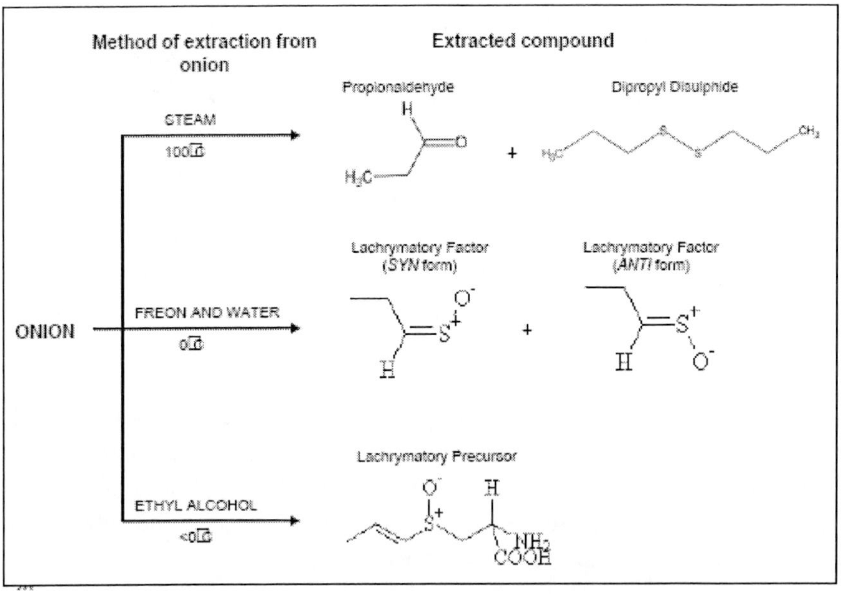

Figure 2. Major organo-sulphur compounds present in different onion preparations based on the extraction method (Taken from Corzo-Martínez *et al.* 2007).

Organo-sulphur compounds present in onion preparations depend on the variety (Yang *et al.*, 2004) and the extraction and/or processing conditions (Figure 2).

Flavonoids, abundant in onion, are also responsible for a great part of the health benefits of these vegetable. In addition, the biological effects of other constituents of intact onion, such as lectins (the most abundant proteins), prostaglandins, fructan, pectin, adenosine, vitamins B_1, B_2, B_6, C and E, biotin, nicotinic acid, fatty acids, glycolipids, phospholipids and essential amino acids, have been studied for over several decades (Fenwick and Hanley, 1985). The importance of biological and pharmacological activities, such as antifungal, antibacterial, antitumor, anti-inflammatory, antithrombotic and hypocholesterolemic properties of certain steroid saponins and sapogenins, such as β-chlorogenin, has been recently demonstrated (Matsuura, 2001; Lanzotti, 2006).

Given the importance of this vegetable as much in food preparation as in medicine, in this chapter the bioactivity of onions has been reviewed, indicating the main responsible compounds.

ANTIMICROBIAL ACTIVITY

In folk medicine, onion has been used for centuries in several societies against fungal, bacterial and viral infections. Due to the great antimicrobial activity that onion possesses, this vegetable could be used like natural preservative in foods to control the microbial growth (Pszczola, 2002).

Recent chemical characterization of their sulphur compounds has allowed stating that they are the main active antimicrobial agents (Rose *et al.,* 2005). However, the application of onion volatile compounds seems to be limited due to their strong flavour, pungent properties and relative biochemical instability. Consequently, there is a growing interest in studying the antimicrobial properties of phenolic compounds which are presumably more stable. Moreover, some proteins and saponins can also contribute to this activity (Griffiths *et al.,* 2002). In the case of onion phenolic compounds, a number of papers below reported have been also addressed not only on the antimicrobial properties but also on the antioxidant activity.

Antibacterial Activity

The major active antibacterial components *in vivo* are the allicin-derived organo-sulphur compounds, such as DAS, DADS (Tsao and Yin, 2001). Several studies have revised the impact of organosulphur-containing compounds on the growth of pathogen by both well diffusion assay tests and minimal inhibitory concentration. It has been found that they present high inhibition against gram positive bacteria of genera *Bacillus*, *Micrococcus*, *Staphylococcus*, *Streptococcus* as well as gram negative bacteria such as *Salmonella enteritidis* or some strains of *Escherichia coli* (Corzo-Martínez *et al*., 2007) and Škerget *et al*., (2009). In addition to organo-sulphur compounds, it has been reported that certain quercetin oxidation products found in onion also present antibacterial activity against *Helicobacter pylori* and MRSA (multidrog-resistant *Staphylococcus aureus*). Additionally, phloroglucinol-3,4-dihydroxybenzoate, quercetin, syringaresinol, and 4-*O*-methylquercetin showed a weak effect against MRSA and a mild effect against *H. pylori* (Ramos *et al*., 2006). Santas *et al*. (2010) studied the antimicrobial activity of flavonol standards and ethyl acetate subfractions of methanolic extracts of three Spanish onion varieties against *Bacillus cereus*, *Staphylococcus aureus*, *Micrococcus luteus*, *Listeria monocytogenes*, *Escherichia coli* and

Pseudomonas aeruginosa. These authors found that, among the onion extracts tested, only ethyl acetate subfraction showed microbial inhibition.

Benkeblia (2004) measured the antimicrobial activity of onion and reported that the essential oil extract had a marked antibacterial activity of certain pathogens, including *Staphylococcus aureus* and *Salmonella enteritidis*. Lately, the addition of dehydrated onion into fresh pork meat cuts has been used to inhibit the growth of total bacteria and Enterobacteriaceae (Park *et al.*, 2008). Very recently, the antimicrobial effects of an onion peel extract obtained by subcritical water (SWE) against *Staphylococcus aureus* have been studied (Lee *et al.*, 2011a). These authors have indicated that the effect of the SWE extract appear to be less effective than quercetin at a similar concentration. In spite of this, they proposed this extract as an adequate additive for the pharmaceutical industry.

In a research on the combination of divergicin M35, a bacteriocin produced by *Carnobacterium divergens* strain M35, and an aqueous extract of onion, among other vegetables, Zouhir *et al.* (2008) stated that, the bactericidal effect of onion extract against *Listeria monocytogenes* appears to be lost when it was combined with divergicin M35, probably due to an antagonist effect between both. Saxena *et al.* (2010) reported the synthesis of silver nanoparticles (Figure 3) by using onion extract and demonstrated that these nanoparticles at a concentration of 50 μg/mL presented a complete antibacterial activity against *Escherichia coli* and *Salmonella typhimurium*.

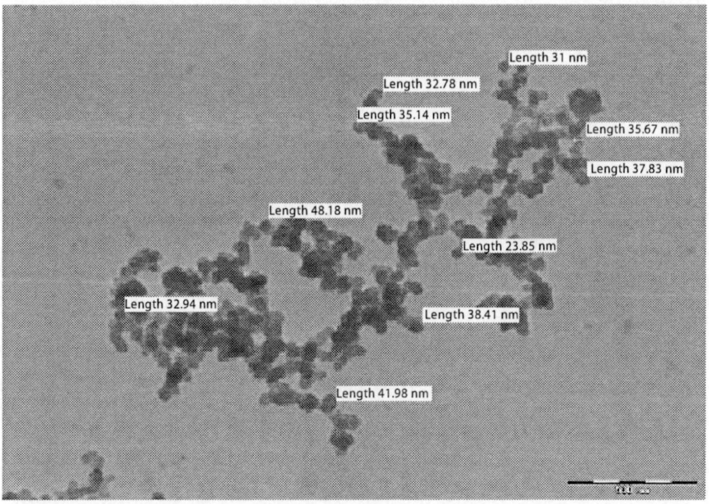

Figure 3. Transmission Electron Microscopy (TEM) micrographs of silver nanoparticles synthesised from onion extract (Taken from Saxena *et al.* 2010).

Antiviral Activity

In comparison to the antibacterial action of onion, hardly any work has been done to investigate its antiviral properties. In addition to sulphur compounds, it has been reported that quercetin (3,5,7,3',4'-tetrahydroxyflavone), the major onion flavonoid, also possesses antiviral activity and enhances the bioavailability of some antiviral drugs (Wu *et al.*, 2005). Onion lectins, unlike the garlic lectins, have a pronounced anti-HIV activity (Van Damme *et al.*, 1993). Additionally, Goren *et al.*, (2002) found a novel medicinal extract derived from onion with broad antiviral activity. This extract may be used to treat or prevent a variety of viral human and animal infections. Examples thereof include retroviral infections such as AIDS, herpes (genital, rectal, oral), distemper, papillomavirus, flu associated influenza viruses, parvoviruses, rhabdoviruses, Epstein Barr virus, CMV, hepatitis virus, RSV, rhinoviruses, and foot and mouth disease virus.

Very recently, Chen *et al.* (2011) have investigated the *in vitro* anti-adenoviral activity of onions, among other *Allium* plants such as shallots, garlic, leeks and green onions, and they have found that shallots present the highest antiviral activity for both ADV41 and ADV3, followed by garlic and onions. According to this and, given the high content of quercetin in onion, even more than in garlic, is it suspected that other phytochemical present in *Allium* plants different from quercetin and its derivatives, could exert a complementary effect on the antiviral action.

Antifungal Activity

The active compounds of onion destroy fungal cells decreasing the oxygen uptake, reducing cellular growth, inhibiting the synthesis of lipids, proteins and nucleic acids, changing the lipid profile of the cell membrane (Ghannoum, 1988) and inhibiting the synthesis of the fungal cell wall (Gupta and Porter, 2001).

Like for the antibacterial activity, the main active antifungal agents from onion extracts are diallyl trisulphide (DATS), DADS and DAS (Tansey and Appleton, 1975). Recently, Borjiham *et al.* (2010) found that Zwiebelane A (cis-2,3-dimethyl-5,6-dithiabicyclo [2.1.1] hexane 5-oxide), a natural product of onion bulbs, enhances the potential fungicidal activity of the typical bactericidal antibiotic Polymyxin B.

Onion extracts are effective against many yeasts species and its essential oil inhibits the dermatophytic fungi (Zohri *et al.*, 1995). Irkin and Korukluoglu (2007; 2009) investigated the antifungal activities of ethyl alcohol or acetone extracts of dehydrated onion against *Aspergillus niger*, *Fusarium oxysporum*, *Candida albicans* ATCC 10231 and *Metschnikowia fructicola.*

De Souza *et al.* (2010) related the levels of total phenolic in onion with the antifungal activity tested against *Rhyzopus oryzae*. The phenolic compounds of onion were extracted in three solvent systems: aqueous, methanolic and ethyl acetate. Among the three systems studied, onion methanolic and aceto-ethylic extracts inhibited efficiently the development of *Rhyzopus oryzae.*

In addition to sulphur compounds, a great variety of antifungal proteins and peptides have been isolated from several *Allium* species, such as the peptide *Ace*-AMP1 obtained from onion seeds with sequence similarity to plant lipid transfer proteins (Phillippe *et al.*, 1995), and allicepin, a novel isolated antifungal peptide from onion bulbs (Wang and Ng, 2004). Wu *et al.* (2011) have carried out a study in which *Ace*-AMP1 was highly expressed in a prokaryotic *Escherichia coli* system as a fusion protein. The purified protein inhibited the growth of many plant fungal pathogens, especially *Alternaria solani*, *Fusarium oxysporum* f. sp. *Vasinfectum* and *Verticilium dahliae.*

Antiparasitic Activity

Regarding the activity that onion and their constituents exert on parasitic protozoa, only a few reports have been published. Due to the occurrence of unpleasant side effects and increasing resistance to the synthetic pharmaceuticals recommended for the treatment of giardiasis, there has been an increasing interest to explore natural alternatives. Antiparasitic properties of onion extracts towards different strains of *Leishmania* and *Trichomonas vaginalis* have been reported as well (Saleheen *et al.*, 2004; Taran *et al.*, 2006).

ANTIOXIDANT ACTIVITY

Oxidation of DNA, proteins and lipids by reactive oxygen species (ROS) plays an important role in aging and in a wide range of common diseases, including cancer and cardiovascular, inflammatory and neurodegenerative diseases, such as Alzheimer's disease and other age-related degenerative conditions (Borek, 1997; Gutteridge, 1993; Richardson, 1993). Research studies evidence that plant-based diets, in particular those rich in vegetable and fruits, provide a great amount of antioxidant phytochemicals, such as vitamins C and E, glutathione, phenolic compounds (flavonoids) and vegetable pigments, which offer protection against cellular damage (Dimitrios, 2006).

As it is known, Onion contains anthocyanins and the flavonoids quercetin and kaempferol. However, anthocyanin pigments, concentrated in the outer shell of red onions, are only minor constituents of the edible portion. Kaempferol, while detectable in certain onion varieties, is present in much smaller amount than quercetin (Bora and Sharma, 2009). Thus, quercetin is the major flavonoid found in onion, present in conjugated form, as quercetin 4'-O-β-glycopyranoside, quercetin 3,4'-O-β-diglycopyranoside, and quercetin 3,7,4'-O-β--triglycopyranoside (Sellappan and Akoh, 2002). The dry outer layers of onion, which are wasted before food processing such as cooking, contain large amounts of quercetin, quercetin glycoside and their oxidative products (Gülsen *et al.*, 2007, Bora and Sharma, 2009), which are effective antioxidants against non-enzymatic lipid peroxidation and oxidation of low density lipoproteins (LDL). Quercetin and its dimerized compound show the highest antioxidative activity, which is comparable to that of α-tocopherol. Therefore, the outer layer extract of onion is expected to be a resource for food ingredients (Ly *et al.,* 2005; Park *et al.*, 2007).

With regards to quercetin bioavailability, Hollman *et al.* (1995) showed that quercetin was indeed absorbed in humans. Recently, it has been demonstrated that its absorption is low in contrast to other dietary antioxidants such as vitamins C and E, limiting its capability to act as antioxidant in plasma *in vivo* (Lotito and Frei, 2006). However, very high interindividual variability has been observed in several studies (Graefe *et al.*, 2001; Moon *et al.,* 2000). Thus, some individuals could absorb quercetin better than others, possibly because of particular polymorphism for intestinal enzymes or transporters.

The glycosides of quercetin are more efficiently absorbed than quercetin itself (Erlund *et al.,* 2000; Graefe *et al.*, 2001) and the nature of the sugar residues in the glycosides influences the extent of absorption. Quercetin is not present in blood as an aglycone but only in conjugated forms. Quercetin-3-*O*-glucuronide, 3'-*O*-methylquercetin-3-*O*-glucuronide, and quercetin-3'-*O*-sulfate have been identified as the major conjugates (Day *et al.,* 2001). Very recently, Jan *et al.* (2010), in a paper on the health benefits of dietary flavonoid quercetin, reviewed the main routes of quercetin glycosides in the major compartments of the gastro-intestinal tract.

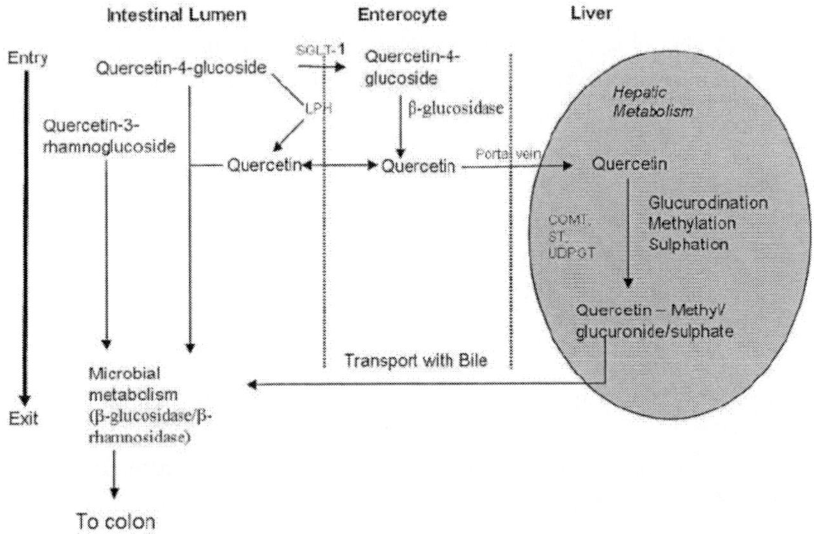

Figure 4. Simplified scheme of the route of quercetin in small intestine (Taken from Jan *et al.* 2010).

According to these authors two different mechanisms could be involved to the intestinal absorption of quercetin and its glycosides (Figure 4). First route facilitating absorption of quercetin glycosides involves luminal deglycosylation by lactate phlorizin hydrolase (LPH) situated in the apical membrane of the small intestine. However, the second way appears to involve sodium dependent glucose transporter-1 (SGLT-1).

Santas *et al.* (2010) confirmed the presence of flavonoids in crude onion methanolic extracts and their activity as antioxidant compounds following the Trolox Equivalent Antioxidant Capacity (TEAC) method. Gökçe *et al.* (2010) determined the antioxidant capacities of a wide range of onion cultivars by means of the "ferric reducing ability of plasma" (FRAP) and TEAC and they suggested that the red onions had higher antioxidant activities than yellow and white onions although yellow onions had the richest phenolic contents. Park and Chin (2010) studied the addition to pork patties of onion extracts (water extract from fresh onion, methanol extract from heated onion and their combinations) and they found that the combination of extracts (1%) had antioxidant activities as effective as butylated hydroxytoluene (0.01%).

Another source of antioxidants can be found in onion waste and by-products, the extraction recovery being one of the most important aspects to be considered. Khiari *et al.* (2009) carried out an investigation on the recovery of antioxidant phenolics from onion solid wastes, composed of the apical trimmings and the outer dry layers of the bulb, employing acidified water/ethanol-based solvent systems. The results indicated the best extraction yields at 6 h whereas the increase of temperature from 40 to 60°C had a negative effect. Singh *et al.* (2009) determined the antioxidant activity by several methods in five extract of red onion peel and they found large amounts of polyphenols in the ethyl acetate (EA) extract, this fraction having a great potential as natural antioxidant in nutraceutical preparations. Benítez *et al.* (2011) also indicated that brown skin and top-bottom of industrial onion wastes could be used as functional ingredient due to their high content in total phenolics and flavonoids with high antioxidant activity, among other bioactive compounds.

Roldán *et al.* (2008) characterized by-products (juice, paste and bagasse) derived from two Spanish onion cultivars ("Recas" and "Figueres") that have been stabilized by thermal treatments and they found that processing of "Recas" onion wastes to obtain a paste and applying a mild pasteurisation were the best alternatives to obtain an interesting stabilised onion by-product with good antioxidant activities measured by DPPH method.

According to Lee *et al.* (2007) onion increases its physiologically active compounds after heating, since they demonstrated that the antioxidant activities of the ethyl acetate fraction were higher in heated (120, 130 and 140 °C) than in raw onion and the higher the temperature of the heat treatment, the greater radical and nitrite scavenging activities. Similar results were found by Woo *et al.* (2007) and they indicated that the optimal heating time and temperature were 2 h and 130°C. Roy *et al.* (2007) analyzed the onion water soluble extracts subjected to thermal treatment at 75 or 100°C for 30 and 60 min and they not only increased the total antioxidant activity but also reduced the pro-oxidant elements. Pérez-Gregorio *et al.* (2011) tried to elucidate the effect of freeze-drying process and storage on onion flavonoids content and demonstrated that the storage of onion powder at room temperature, in dark, in air- and water-tight glass bottles for up to 6 months was keeping rather stable all antioxidant flavonoids.

In addition to the afore-mentioned compounds, other identified antioxidant compound is N-fructosyl lysine, Amadori rearrangement product, originated during the first steps of the Maillard reaction as a result of the processing and storage, mainly to high temperatures, has been studied. Moreno *et al.* (2006) determined by $ORAC_{FL}$ assay the evolution of the antioxidant activity (AA) of dehydrated onion stored at 0.44 a_w and 30 and 50 °C, and they found an increase in AA in agreement with the Maillard reaction evolution. These authors suggested that, although the Amadori compounds could exert a moderate effect on the AA, the advanced reaction products are the major contributors to this property.

ANTICARCINOGENIC AND ANTIMUTAGENIC ACTIVITIES

Many epidemiologic and *in vitro* and *in vivo* laboratory studies have been developed to evidence the chemopreventive or anticarcinogen effects of onion and related *Allium* species (Bianchini and Vainio, 2001; Galeone *et al.* 2006; Roldán-Marín, 2009). In general, these studies are more consistent in reporting a protective effect of onion in gastric cancer, an inversely correlation between onion intake and the risk of the stomach cancer being observed (Dorant *et al.,* 1996; You *et al.*, 1998; Gao *et al.*, 1999; Hsing *et al.*, 2002; Gonzalez *et al.,* 2006; Kim and Kwon, 2009; Bang and Kim, 2010). The chemopreventive effects of onion against stomach and esophagus cancers may be related to their antibacterial properties. Inhibition of bacterial growth in the gastric cavity may result in less conversion of nitrate to nitrite in the stomach, a decreased probability of endogenous formation of carcinogenic N-nitroso compounds, and reduction in *H. pylori* infection specifically (Dorant *et al.,* 1996).

Onion intake has been also consistently associated with a decreased risk of colorectal (Steinmetz *et al.,*1994; Millen *et al.,* 2007; Taché *et al.,* 2007), lung (Sankaranarayanan *et al.,* 1994; Dorant *et al.*, 1996; Le Marchand *et al.,* 2000), brain ((Hu *et al.,* 1999), prostate (Hsing *et al.* 2002), bladder (Malaveille *et al.*, 1996), liver (Fukushima *et al.,* 2001), breast (Levi *et al.,* 1993; Challier *et al.*, 1988), ovarian (Shen *et al.*, 1999), endometrial (Galeone *et al.*, 2009a), and skin (Byun *et al.*, 2010) cancers.

These effects appear to be mediated by various mechanisms, which are not fully understood. On the basis of several studies, it is possible to state that mechanisms by which onion exerts their anticarcinogenic and antimutagenic action include (Figure 5): alteration of carcinogen metabolism by inducing phase II enzymes such as glutathione S-transferase (GST), NAD(P)H-dependent quinine reductase, and UDP-glucuronosyl transferase (Tsuda *et al.,* 2004), that increase the carcinogen polarity, facilitating its excretion from the body (Guyonnet *et al.,* 2001; Brisdelli et al., 2007); inhibition of bioactivating enzymes of procarcinogens (Lautraite *et al.,* 2002; Muto *et al.,* 2001; Platt *et al.,* 2010); inhibition of oxidative damage due to its antioxidant action (Perchellet *et al.*, 1990; Mutoh *et al.*, 2000; Raso *et al.,* 2001).

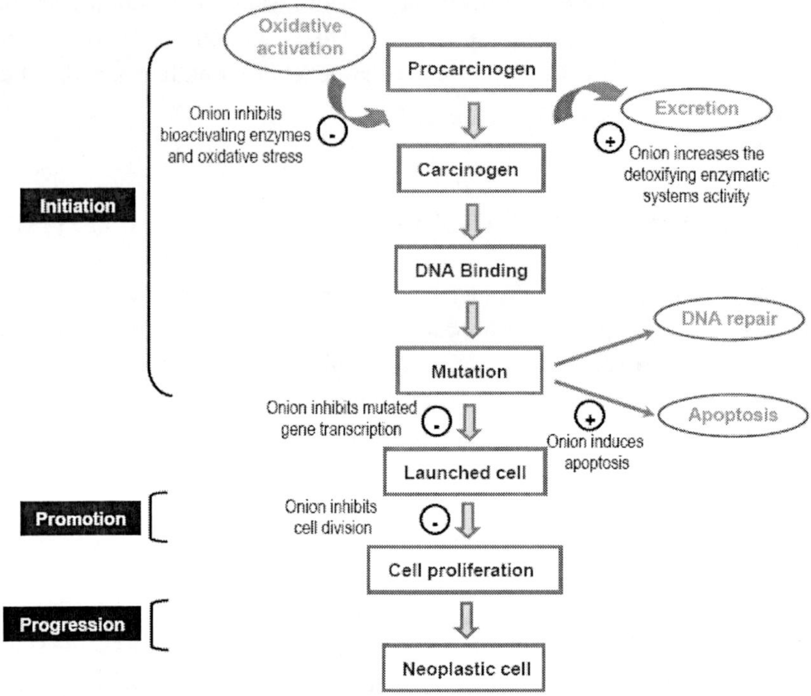

Figure 5. Cholesterol biosynthesis pathway (Taken from Cardelle-Cobas *et al.* 2009b).

Recently, Bang and Kim (2010), in a study on the effect of onion on the chemical induction of preneoplastic lesions in rat liver, reported that onion inhibits early-stage hepatocellular carcinogenesis through the suppression of oxidative stress by modulating the GST and glutathione peroxidase activity; inhibition of cellular proliferation by induction of apoptosis and inhibition of cell division (Perchellet *et al.*, 1990; Brisdelli *et al.*, 2007); gene transcription inhibition (Miodini *et al.,* 1999; Bora and Sharma, 2009); protection against UV-induced immunosuppression (Steerenberg *et al.,* 1998; Bora and Sharma, 2009); and inhibition the lipoxygenase and cyclooxygenase activities (anti-inflammatory effect) (Perchellet *et al.*, 1990; Mutoh *et al.*, 2000; Raso *et al.*, 2001; Rose *et al.*, 2005).

Regarding bioactive compounds, several investigations have shown that both water- and lipid-soluble sulphur compounds from onion provide, at least in part, its anticarcinogenic activity. Among them, DAS, diallyl disulphide (DDS), dipropyl sulphide (DPS), dipropyl disulphide (DPDS), N-acetylcysteine, S-allyl cysteine (SAC), and S-methylcysteine (SMC) have shown to inhibit both early and late stages of colon, forestomach, esophagus, mammary gland, lung, liver and kidney carcinogenesis (Reddy *et al.,* 1993; Takada *et al.,* 1997; Fukushima *et al.,* 1997; Guyonnet *et al.*, 2001; Fukushima *et al.*, 2001; Bora and Sharma, 2009).

Another sulphur compounds, as methiin (Takada *et al.*, 1997), inhibits the cellular proliferation by inducing apoptosis in human cell cultures, like, for example, in human leukaemic cells. More recently, it has been postulated that organo-sulphur compounds such as tetrasulfides ocurring naturally in onion are able to suppress the proliferation of sensitive and resistant human breast carcinoma cells by targeting the cell division cycle 25 phosphatases, crucial enzymes of the cell cycle (Viry *et al.*, 2011)

In addition to organo-sulphur compounds, organo-selenium compounds are largely responsible for the anticarcinogenic activity of onion (Matsuura, 1997; El-Bayoumy *et al.,* 2006). Thus, it has been observed that Se-enriched onion has higher anticarcinogenic activity than the common plants (Ip *et al*, 1992). This increased effect of cancer prevention is achieved at least partly by S substitution with Se.

The pure Se-compounds have proved to be superior anticancer agents than their corresponding S-analogues. For example, diallyl selenide is at least 300 times more active than DAS in the reduction of tumours of mammal cancer (El-Bayoumi *et al.*, 1996). The two major Se-compounds possessing anticancer activity in onion are γ-glutamyl-Se-methyl selenocysteine (Finley, 2005; Hurst *et al.* 2010) and Se-methyl selenocysteine, this latter being the most chemopreventive (Block *et al*, 2001). Other forms of selenium identified in uncooked onions include selenomethionine, selenocysteine, and selenite/selenate (Kotrebai *et al.,* 2000).

Anticarcinogenic and antimutagenic properties of onion may be also partly attributed to its abundance of phenolics, including flavonoids. Several recent studies (Jang and Lim, 2009; Jeong *et al.,* 2009) have reported the different anticancer activity of extracts from flesh and peel of white, yellow and red onion as a function of their total phenolics and flavonoids, as quercetin, level. In general, onion peel, with the highest amounts of total phenolics and flavonoids, inhibited the growth of several human cancer cell lines, including cells of stomach, colon (Jang and Lim, 2009), breast, and prostate cancer (Jeong *et al.*, 2009), more efficiently than onion flesh. In addition, extracts from white onion was less effective than those from yellow and red onion, with a higher content in total phenolics and flavonoids than the former.

Likewise, in a study on the potential antioxidant and antimutagenic activities of several extracts with different polarity from red onion peel, Singh *et al.* (2009) suggested that the large amount of polyphenols contained in the ethyl acetate fraction, including ferulic, gallic, protocatechuic acids, quercetin and kaempferol, might be the cause of their strong antioxidant and antimutagenic properties.

Particularly, quercetin and derivatives exhibit anticancer properties, which have been demonstrated in a number of malignancies, including prostate, breast, skin, lung and liver cancers (Avila *et al.,* 1994; Musonda and Chipman, 1998; Le Marchand *et al.*, 2000; Le Marchand, 2002; Vijayababu *et al.*, 2006; Arung, Furuta, Ishikawa, Kusuma, Shimizu, and Kondo, 2011). Likewise, several studies have reported that quercetin enhances bioavailability of some anticancer drugs, as Tamoxifen, a non-steroidal antiestrogen for the treatment and prevention of breast cancer, by promoting their intestinal absorption and reducing their metabolism (Shin *et al.*, 2006; Wu, *et al,* 2005).

In addition, a recent study has reported that the combined effect of quercetin and sulforaphane [1-isothiocyanato-4-(methylsulfinyl)-butane] (a member of an isothiocyanate family of chemopreventive agents isolated from broccoli) on the proliferation and migration of melanoma (B16F10) cells is more effective than either compound used alone (Pradhan *et al.*, 2010).

Besides quercetin, luteolin (3,4,5,7-tetrahydroxyflavone), a natural flavonoid abundant in onions, has shown antiproliferative (Huang *et al.*, 1999), antimetastatic (Huang *et al.,* 1999; Lee *et al.,* 2006), antioxidative (Manju *et al.*, 2005), antiangiogenic (Bagli *et al.*, 2004), and anti-inflammatory (Ueda *et al.*, 2002) effects, primarily in cancer cell assay models. One study showed that luteolin inhibits chemically induced skin tumorigenesis in a mouse model (Ueda *et al.*, 2003). Recently, Byun *et al.* (2010) reported that luteolin exerts its protective effect against UVB-induced skin tumorigenesis in SKH-1 hairless mice by directly suppressing PKCε and c-Src kinase activity, two protein kinases closely associated with the development of UV-induced skin cancer.

Together with all these bioactive compounds, new chemicals are being isolated from onion extracts and characterized. Among them, 2,3-dihydro-3,5-dihydroxy-6-methyl-4H-pyranone and 5-hydroxy-3-methyl-4-propylsulfanyl-5H-furan-2-one have shown to prevent or inhibit cancer cell growth *in vitro* by inducing apoptotic cell death through the inhibition of NF-κB (Ban *et al.,* 2007) and by increasing the quinone reductase activity, a phase II xenobiotic metabolizing enzyme (Xiao and Parkin, 2007), respectively.

CARDIOVASCULAR PROTECTIVE EFFECTS

Cardiovascular diseases (CVD) include coronary heart disease (heart attacks), cerebrovascular disease, raised blood pressure (hypertension), peripheral artery disease, rheumatic heart disease, congenital heart disease and heart failure. If current trends are allowed to continue, it has been estimated that about 20 million people will die from CVD (mainly from heart attacks and strokes) (WHO, Cardiovascular disease) by 2015. Therefore, CVD have a major impact on the mortality and quality of life of human populations across the world, despite improvements in lifestyle and innovations in the prevention and treatment of CVD in previous decades (Wensing *et al.*, 2009; Roldán-Marín, 2009).

There are many factors associated with arteriosclerosis and cardiovascular diseases, among which can be included: elevated blood cholesterol and triglycerides levels, including LDL-cholesterol; increased platelet activity, which can give rise to arteriosclerotic plaques formation; elevated blood homocysteine; diabetes; hypertension; and obesity. These cardiovascular disease risk factors are mainly determined by uncontrollable (heredity, gender and age) and lifestyle-related causes (smoking, physical inactivity, stress and unhealthy diet), which are possible to be modified. For this reason, a potential approach to the prevention and treatment of CDV could be based on the diet. In this sense, onion has been described to have hypolipidemic, hypoglycemic, and antithrombotic effects and, therefore, could be useful in a CVD preventive diet, according to the study carried out by Galeone *et al.* (2009b), the first from Mediterranean countries.

Effects on Levels of Serum Lipids

The synthesis and utilization of cholesterol must be tightly regulated in order to prevent over-accumulation and abnormal deposition within the body, since the abnormal deposition of cholesterol and cholesterol-rich lipoproteins in the coronary arteries eventually leads to atherosclerosis.

Slightly less than half of the cholesterol in the body derives from biosynthesis *de novo*. Biosynthesis in the liver accounts for approximately 10%, and in the intestines approximately 15% of the amount produced each day. Cholesterol synthesis occurs in the cytoplasm and microsomes from the two-carbon acetate group of acetyl-CoA (King and Marchesini, 2007), as shown in Figure 6.

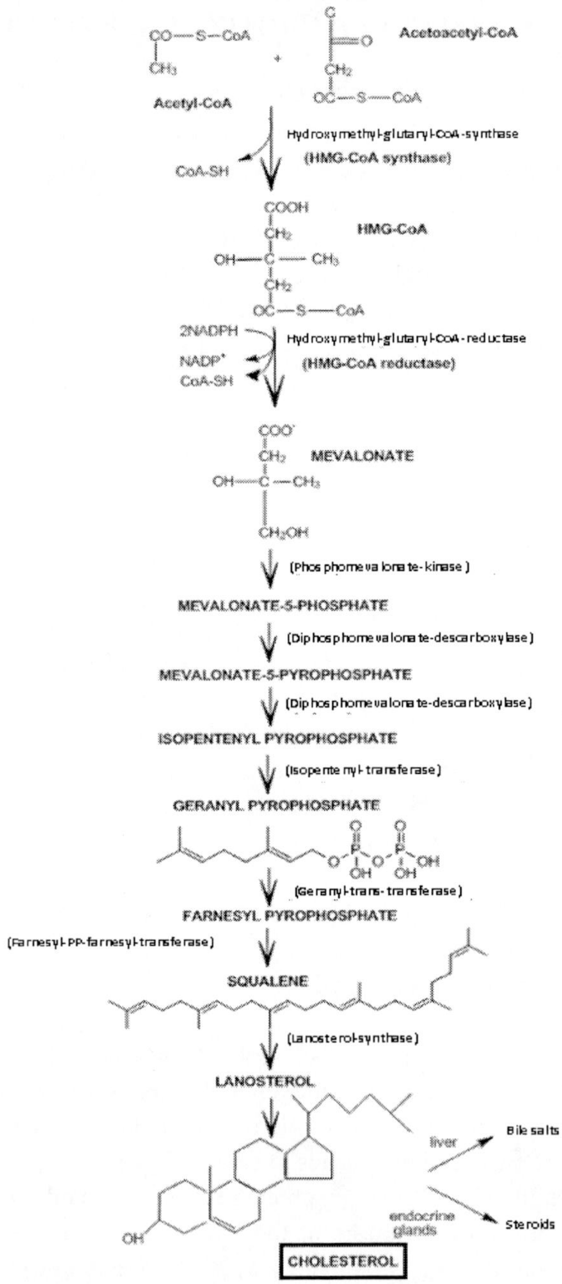

Figure 6. Modes of action by which onion and its derivatives exert their anticarcinogenic activity.

Onion has been reported to reduce moderately blood triglycerides levels and to inhibit hepatic cholesterol biosynthesis in experimental animal such as healthy rats, rabbits and pigs fed a high fat diet (Sharma *et al.,* 1975; Vatsala *et al.,* 1980; Lata *et al.*, 1991; Effendy *et al.*, 1997; Ostrowska *et al.*, 2004; Gabler *et al.*, 2006; Roldán-Marín *et al.*, 2010), no significant differences existing between hypolipidemic and hypocholesterolemic effects of onion and garlic (Emmanuel and James, 2011).

Studies with humans have been also carried out. Thus, a group of volunteers fed a high fat diet plus 100 g onion once a day and those fed fat diet only showed a significant decrease in serum triglycerides, but not cholesterol, as compared to those only fed with fat diet only (Sainani *et al.*, 1978). Another study reported that oral administration of a butanol onion extract to patients with elementary lipemia prevented an increase in total serum cholesterol, J-lipoprotein cholesterol, and J-lipoprotein and serum triglycerides (Jain and Vyas, 1977). Similarly, a saponin fraction (50 mg) and the bulb (100 mg) of onion have also shown to decrease serum cholesterol and plasma fibrinogen levels (Dorsch and Wagner, 1991). Moreover, a recent study indicated that intake of onion concentrated extracts exerts beneficial effects on dyslipidemia by reducing serum total cholesterol and LDL-cholesterol levels in borderline hypercholesterolemic subjects (Lee *et al.,* 2010). All these studies showed that onion intake may to inhibit the formation of atherosclerotic plaques and, consequently, to reduce risk indices of CVD.

Among bioactive compounds involved in onion hypolipidemic and hypocholesterolemic effects, organo-sulphur compounds are the main active, as much in humans as in experimental animals (Yeh *et al.,* 1997; Liu and Yeh, 2002). Volatile oil of onion and S-methyl cysteine sulphoxide (SMCS) have shown to possess the ability of counteract the lipogenic effect of sucrose, alcohol and cholesterol diets (Wilcox *et al.,* 1984; Farya *et al.,* 1986; Kumari and Augusti, 2007; Bora and Sharma, 2009).

Mechanisms of action by which these onion bioactive compounds exert their hypolipidemic and hypocholesterolemic activities include: inhibition hepatic lipid/cholesterol biosynthesis by inactivating thiol enzymes (eg. HMG-CoA), which promote it, or by reducing the level of NADPH in tissue, thus they may not be available for cholesterol synthesis (Gebhardt *et al.*, 1994; Kumari and Mathew, 1995; Gupta and Porter, 2001); and enhancement of cholesterol turnover to bile acids and its excretion through gastrointestinal tract (Srinivasan and Sambaiah, 1991).

In addition to organo-sulphur compounds, flavonoid quercetin and derivatives have also shown to be able to reduce serum concentrations of total cholesterol and LDL-cholesterol, and to increase serum concentrations of HDL-cholesterol (Glasser *et al.,* 2002; Lee *et al.*, 2011b). Moreover, Terao *et al.* (2008) reported that quercetin metabolites are incorporated into the atherosclerotic region and act as complementary antioxidants, when oxidative stress is loaded in the vascular system.

Conversely to the above mentioned works, one study reported no significant changes in cholesterol or lipid levels of the eye in rabbits, after treatment of the animals for six months with an aqueous onion extract (20% of diet) (WHO monographs, 1999). Similarly, Sharma and Sharma (1976; 1979) observed that fresh onion extract (50 g) did not produce any significant effects on serum cholesterol, fibrinogen or fibrinolytic activity in normal subjects. Therefore, although onion appears to hold promise in reducing parameters associated with cardiovascular disease, more in-depth investigations are required. Moreover, it is important to note that onion is mostly consumed after processing rather than raw which can lead to a certain decrease in its content of bioactive compounds. In this respect, Gorinstein *et al.* (2010) recently reported that blanching of onion for 90 s most fully preserves the contents of its bioactive compounds and related antioxidant potential. Thus, diets, supplemented with red onion and to a lesser degree with white onion, significantly hindered the rise in plasma lipids levels and the decrease in the plasma antioxidant activity in cholesterol-fed rats.

Hypotensive and Bradycardic Effects

Epidemiological studies have demonstrated that elevated blood pressure is one of the major risk factors for stroke and coronary heart disease. A close association between blood pressure and the incidence of cardiovascular diseases is well established if systolic/diastolic blood pressure is above 140/90 mmHg. In recent years, popular blood pressure-lowering nutraceuticals and functional foods, including onion, have attracted considerable interest as potential alternative therapies for treatment of hypertension, especially for pre-hypertensive patients, whose blood pressure is marginally or mildly high but not high enough to warrant the prescription of blood pressure-lowering medications (Chen *et al.,* 2009).

Onion has been shown to be anti-hypertensive in many *in vivo* animal studies. In L-NAME (NG-nitro-L-arginine methyl ester)-induced hypertensive rats and stroke-prone spontaneously hypertensive rats (SHRSP), dried onion was able to reduce blood pressure when it was added into diet at 5% (Sakai *et al.,* 2003). In addition, allylmercaptocaptopril (CPSSA), a drug assayed in hypertensive rats and synthesised through the reaction of the pharmaceutical drug Captopril with allicin, provides better protection against hypertension. This is due to it has the Captopril ability to inhibit the angiotensin-converting enzyme (ACE) and the allicin ability to reduce serum cholesterol and triglycerides levels (Miron *et al.*, 2004).

Although the active ingredients responsible for the blood pressure-lowering activity of onion are not yet fully understood, some evidence suggests that they reduce blood pressure probably by the following mechanisms: (i) by increasing the level of nitric oxide (NO) and the activity of nitric oxide synthetase (NOS). This has been observed in studies with SHRSP (Sakai *et al.*, 2003) and, recently, with cultured human umbilical vein endothelium cells (Jiemei *et al.*, 2011). It has also been shown that phenolics, flavonoids and 3-mercapto-2-methylpentan-1-ol (3-MP) of onion were able to scavenge the peroxynitrite radical *in vitro,* inhibiting, thus, peroxynitrite-induced nitration of protein tyrosine residues, considered as one of the major pathological causes of several human diseases, including cardiovascular disorders (Rose *et al.,* 2003; Ho *et al.,* 2010); (ii) by inhibiting the production of angiotensin II. A study with several rat models of hypertension has indicated that quercetin and its methylated metabolite isorhamnetin, found in onion, can reduce blood pressure and prevent angiotensin II-induced endothelial dysfunction by means of the inhibition of the overexpression of p47 (phox), a regulatory subunit of the membrane NADPH oxidase, and the subsequent increased superoxide production, resulting in a highest NO bioavailability (Sanchez *et al.*, 2007); and (iii) through the inhibition of calcium influx (Naseri *et al.,* 2008a).

Regarding the effect of onion on blood pressure in humans, available data are scarce. Mayer *et al.* (2001) conducted a randomized, placebo-controlled, double-blind, and crossover study to investigate the effect of an onion-olive oil maceration capsule formulation on arterial blood pressure. They found that it produced a decrease in arterial blood pressure. Moreover, one study investigated the efficacy of quercetin supplementation on lowering blood pressure in hypertensive humans and demonstrated that 730 mg of quercetin per day could reduce the systolic blood pressure by 7 mmHg, the diastolic blood pressure by 5 mmHg, and mean arterial pressures by 5 mmHg in stage 1 hypertensive patients (Edwards *et al.,* 2007). Similarly, Egert *et al.* (2009) recently found that quercetin reduced systolic blood pressure and plasma oxidized LDL concentrations in overweight subjects with a high-CVD risk phenotype in a double-blinded, randomised, placebo-controlled cross-over trial.

Anti-Hyperglycemic or Anti-Diabetic Potential

The relationship between diabetes Mellitus and atherosclerosis is likely based on the interactions between arterial cells and atherogenic glycosylated LDL lipoproteins originated during diabetes development. These play a key role in the initiation of an atherosclerotic lesion, inducing cholesterol accumulation in arterial cells (Ide and Benjamin, 2001) and other more severe atherosclerotic manifestations at cellular level (Winocour, 1994; Sobenin *et al.,* 1994).

The effectiveness of onion and its derivatives as hypoglycemic agents has been shown in several studies either with diabetic animal models or humans (Brahmachari and Augusti, 1962; Augusti, 1973; Jain and Vyas, 1974; Sharma *et al*., 1977; Bever and Zahnd, 1979; Ashwah *et al.,* 1981; Srinivasan, 2005; El-Demerdash *et al.,* 2005). Results of a recent study (Bang *et al.,* 2009) have indicated that this beneficial ameliorating influence of dietary onion on diabetic nephropathy may be mediated through onion's ability to decrease blood glucose, serum lipid/cholesterol levels and lower renal oxidative stress in streptozotocin-induced diabetic rats. Babu and Srinivasan (Babu and Srinivasan, 1997) observed that dietary onion intake for 8 weeks produced significant hypolipidemic effect besides hypoglycemic influence in diabetic rats. More recently, Lee *et al.* (2008) showed that onion peel was effective in controlling hyperglycemia in animal models of type 2 diabetes Mellitus, at least in part by inhibiting alpha-glucosidase activity (Lee *et al.,* 2008).

The organo-sulphur compounds S-methylcysteine sulphoxide (SMCS) and S-allylcysteine sulphoxide (SACS) have been related to significant amelioration of weight loss, hyperglycemia, low liver protein and glycogen, and other characteristics of diabetes Mellitus in rats (Sheela and Augusti, 1995). The use of SMCS and SACS (200 mg/kg/day) gave results comparable to treatment with insulin or glibenclamide but without the negative side effect of cholesterol synthesis stimulation. Antidiabetic effect of SMCS was also reported by Kumari and Mathew (1995). This compound exerts its anti-diabetic action by 3 different ways: (i) stimulating the insulin production and secretion by pancreas, (ii) interfering with dietary glucose absorption, and (iii) favouring the insulin saving (Srinivasan, 2004a; Srinivasan, 2004b).

In addition to organo-sulphur compounds, anti-hyperglycemic and anti-diabetic activities of diphenylamine (Karawya *et al.,* 1984) and quercetin from onion have been also reported. *In vivo* analysis of the effects of quercetin on human diabetic lymphocytes showed a significant increase in the protection against DNA damage from hydrogen peroxide at the tissue level (Lean *et al.,* 1999). Likewise, it has been reported that long-term absorption of quercetin could be useful to prevent advanced glycation of collagens, which contributes to development of cardiovascular complications in diabetic patients (Urios *et al.,* 2007). In addition, a very recent study (Jung *et al.,* 2011) carried out with high fat diet/streptozotocin-induced diabetic rats has shown that onion peel extract, containing a high content in quercetin, might improve glucose response and insulin resistance associated with type 2 diabetes, even with a greater potency than pure quercetin equivalent, by alleviating metabolic dysregulation of free fatty acids, suppressing oxidative stress, up-regulating glucose uptake at peripheral tissues, and/or down-regulating inflammatory gene expression in liver. These findings provide a basis for the use of onion peel to improve insulin insensitivity in type 2 diabetes Mellitus.

Moreover, the use of onion has been suggested in conjunction with anti-diabetic drugs to increase their therapeutic potential and to minimize their oral dosage. According to experimental data, 50 g onion, daily incorporated in diabetic diet, could serve as an effective supportive therapy in the prevention and maintenance of long-term complications of diabetes.

Anti-Platelet or Anti-Thrombotic Effect

The major function of blood platelets is to maintain the haemostatic integrity of blood vessels and to stop bleeding after injury (Ali *et al.*, 2000), through vasoconstriction, clot formation and blood coagulation. However, when there is an imbalance in the blood coagulation system, a blood clot called thrombus can be formed (Figure 7) and block the flow of blood through a vein or artery, and even can detach from the vessel wall to become a life-threatening embolus when it lodges in the lungs or other vital organs. Likewise, blood clots in coronary arteries cause acute coronary syndrome and blood clots that form in the heart are the major cause of stroke in people with atrial fibrillation. Therefore, it is evident that thrombosis complications play a major role in CVD (Becker, 1999).

Inhibition of platelet aggregation by onion has been demonstrated *in vitro* and *in vivo* (Ali, Bordia, and Mustafa, 1999; Ali *et al.,* 2000; Briggs *et al.,* 2001; Jung *et al.,* 2002; Hubbard *et al.,* 2006; Bora and Sharma, 2009). Studies on the antithrombotic action of onion have reported that its aqueous extracts inhibit adenosine diphosphate, collagen, epinephrine and arachidonic acid-induced platelet aggregation *in vitro* (Srivastava, 1984).

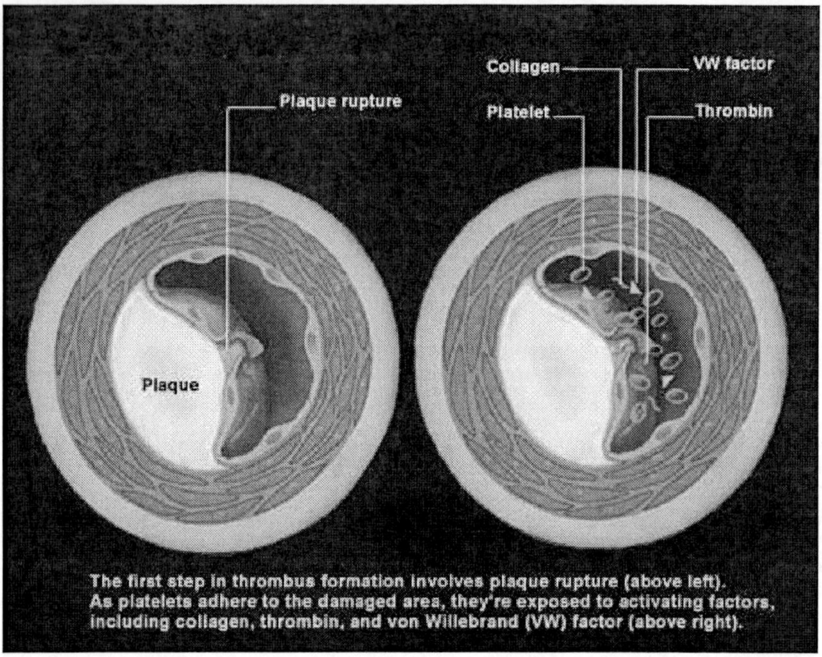

Figure 7. Formation of a clot or thrombus into a blood vessel obstructing the flow of blood through the circulatory system.

Likewise, the essential oil, and butanol and chloroform extracts inhibited platelet aggregation in rabbits (Makheja *et al.,* 1979; Ariga and Oshiba, 1981). Moreover, both raw onion and its essential oil increased fibrinolysis *in ex-vivo* and increased the coagulation time in rabbits (Breu and Dorsch, 1994). Aggregation of human platelets by onion was also inhibited *in vitro* by its essential oil and ethanol, butanol and chloroform extracts (10-60 Kg/ml) (Vanderhoek *et al.,* 1980) through the decrease of thromboxane A2 synthesis, a potent inducer of platelet aggregation (Makheja and Baily, 1990; Moon *et al.,* 2000). *In vivo* effects of onion intake in rats (500 mg/kg) also showed significant inhibition of serum thromboxane A2 (Bordia *et al.*, 1996). Low dose (50 mg/kg) showed little effect, but benefit was observed over long-term consumption. Similarly, raw Welsh onion extracts showed vasodilating effects on precontracted aortic rings of rats as well as to prolong bleeding time, reduce platelet aggregation and increase cAMP level (Chen *et al.,* 1999). Onion, in its raw form, is recognized as an antiplatelet agent. However, as mentioned above, onion is generally cooked before consumption, probably losing its anti-aggregatory effect. Several studies have found that boiled onions, even at the high dosage level, showed no anti-thrombotic effect, probably due to degradation of bioactive compounds. Moreover, extensive heating may result in pro-aggregatory effects (vasoconstriction and induction of thromboxane synthesis). These results suggest that in order to obtain the maximum health benefits, onions should be eaten raw or moderately cooked (Bordia *et al.*, 1996; Chen *et al.*, 1999; Cavagnaro *et al.,* 2007).

On the other hand, Ali *et al.* (1999) showed antiplatelet activity in rabbit plasma, but not in human plasma and suggested that varietal differences may play a role (Ali *et al.*, 1999). In agreement with this, several studies have reported that antiplatelet activity is substantially affected by genotype, environment and storage duration of vegetable. It has been reported that, in onions, the antiplatelet activity is determined, in part, by the native concentration of organo-sulphur compounds and genotypically determined sulphur content of the bulb (Goldman *et al.,* 1996). According to this, a number of epidemiologic studies have reported that antiplatelet activity of onion is considered to be a property of organo-sulphur compounds, concretly thiosulphinates and a class of α-sulphinyl-disulphides (cepaenes) (Breu and Dorsch, 1994; Block *et al.,* 1997). These compounds have structural similarity to ajoene, considered the major antiplatelet compound in garlic extracts.

In addition, other non-sulphur compounds, such as β-chlorogenin and quercetin, have also shown to inhibit platelet aggregation (Rahman *et al.,* 2006). Quercetin and its derivatives exert their beneficial effects on cardiovascular health by antioxidant and anti-inflammatory activities (Kuhlmann *et al.,* 1998), through the inhibition of lipid peroxidation and endothelial cell damage, which are involved in the early development of atherosclerosis (Da Silva *et al.,* 1998; Kaneko and Baba, 1999). An *in vitro* study carried out by Janssen *et al.* (1998) showed that 2500 μmol/L quercetin isolated from onions inhibited platelet aggregation by 95-97%. However, an *in vivo* assay from the same authors with 18 human subjects ingesting 114 mg quercetin/day showed no significant effects. Therefore, it was concluded that necessary concentration levels of quercetin for beneficial effects were too high to be obtained dietarily.

Moreover, it has been recently reported the anti-inflammatory effect of two phenylpropenoic acid amides isolated from *Allium fistulosum* (green onion), called Typheramide (N-caffeoyltyramine) and alfrutamide (Nferuloyltyramine). These compounds have shown to significantly inhibit cyclooxygenases COX 1 and 2, which are principally involved in catalyzing the processing of arachidonic acid to several prostaglandins and thromboxanes (e.g., thromboxane A2, thromboxane B2), promoting the platelet aggregation (Park, 2011).

Other Beneficial Effects

Thiosulphinates and cepaenes derived from onion have been shown to possess antiasthmatic activity (Dorsch, 1996), due to the inhibitory effect of cyclooxygenase and lipoxygenase mediated reactions which initiate eicosanoid metabolism and lead to bronchial restriction, as it has been demonstrated *in vitro* (Wagner *et al.,* 1990). In general, saturated thiosulphinates are less active than unsaturated ones and cepaenes are more active than thiosulphinates. Likely, these effects *in vitro* are responsible, at least in part, for onion extracts anti-inflammatory and antiasthmatic properties observed *in vivo* (Breu and Dorsch, 1994).

The anti-allergic effects of an herbal fraction (ALC-02) derived from the bulb were evaluated in rats by Kaiser *et al.* (2009). Concretely, the efficacy was tested against various events responsible for Type I allergic reactions and the authors attributed the antiallergic profile of ALC-02 to its potential antihistaminic, anti-inflammatory and antioxidant activities.

It has been reported that onion stimulates the digestive process, accelerating digestion and reducing food transit time in the gastrointestinal tract (Platel and Srinivasan, 2001). Naseri *et al.* (2008b) investigated, among other properties, the spasmolytic activity of onion peel powder on rat ileum contractility. Onion peel extracts obtained in 70% alcohol inhibited ileum contractions without involving beta-adrenoceptor, opioid receptor, nitric oxide production and potassium channels activation. According to this, the authors suggested that quercetin in onion peel extracts seems to induce spasmolytic effect via calcium channels.

Onion prebiotic activity has been also investigated (Sharma *et al.*, 2006; Benkeblia and Shiomi, 2006) due to their high soluble fibre content, specially inulin and fructooligosaccharides (FOS) (Cardelle-Cobas *et al.*, 2009a) which stimulate in the colon the growth of specific microorganisms, as bifidobacteria and lactobacilli, with a general positive health effect (Gibson, 1998; Ernst and Feldheim, 2000). Both, inulin and FOS of onion may be used as functional ingredients to enrich many processed foods without any negative impact on their taste (Causey *et al.,* 2000). Roldán-Marín *et al.* (2009) reported the effects of onion extract and onion by-product derived from a pasteurized paste on gut environment in healthy rats and they observed a prebiotic effect of these products as evidecend by decreased pH, increased butyrate production and altered gut microbiota enzyme activity.

Other positive effect of onion which has been assessed in *in vivo* (rats) is the androgenic effects of different doses of onion bulb juice on sperm parameters by using hormone measurements and histopathological studies (Khaki *et al.* 2009).

In this study, freshly prepared onion juice significantly affected the sperm number, percentage of viability, and motility. Thus, 4g/kg of freshly prepared onion juice is effective in sperm health parameters. Moreover, onion intake has shown to be inversely associated with benign prostatic hyperplasia (BPH), disease that involves the formation of large, fairly discrete nodules in the periurethral region of the prostate as a result of an accelerated proliferation of prostate cells due to the influx of androgens (testosterone and related hormones) at high concentrations. This was observed by Galeone *et al.* (2007) in a multicenter case-control study of 1369 patients with BPH and 1451 controls.

CONCLUSION

It can be said that onion is very suitable as food ingredient, not only due to its recognized and appreciated organoleptic properties but also because of the wide range of important biological activities (antimicrobial, antioxidant, anticarcinogenic and antimutagenic, hypolipidemic and hypocholesterolemic, anti-hypertensive, anti-thrombotic, and anti-hyperglycemic activities, prebiotic character, and immunosuppressive, neuroprotective, and anti-inflammatory effects) that it possesses.

To date, many favourable experimental and clinical effects of onion and onion preparations have been reported. These health properties are mainly associated with the following types of chemical compounds: i) non-structural and soluble carbohydrates such as FOS; ii) organo-sulphur compounds, which are also responsible for the pungent aroma and taste; iii) organo-selenium compounds; and iv) phenolic compounds such as flavonoids, particularly quercetin derivatives. In most of the cases these constituents can share different biological activities.

It is well extended the belief that the therapies based on natural remedies have fewer side effects than pharmaceutical products. However, in spite of the numerous studies about the beneficial effects of onion, certain aspects need to be taken into account such as the adverse effect related to bad breath, body odour, gastrointestinal upsets and interactions with other constituents and drugs, although these effects depend on the dosage and susceptibility of the individual. In this regard, the effective dosage to note the beneficial effects as well as the most suitable preparation to avoid undesirable effect should be defined.

Other very important topics are related to processing of onion and its derivatives, since the sensitivity to heat of active components of onion can question the efficacy of different commercial preparations as therapeutic agents, moreover the hard odour and taste of onion extracts can contribute to the reject of these products. For these reasons, more investigations are needed to find a product without odour and taste and that preserves all the biological properties of raw onion.

Clues of vegetable benefits are sometimes found in epidemiological results in which studies of population diet are correlated with the incidence of a particular type of disease. In addition, most of the research works have been done in *in vitro* assays and, in some examples, in *in vivo* by using experimentation animals such as rats.

In these investigations the improved benefits are viewed as positive but whether this translates to a "health benefit" in some cases is unclear. For these reasons, although some assays have been carried out in humans, more clinical studies are necessary to determine if the results with animals can be totally extrapolated to humans.

In this sense, one key question is to know how and in what form the bioactive compound is present in the digestive tract to assure a proper bioavailability and effect. Although some recent papers have studied the bioavailability of some component such as quercetin, more research is needed, particularly in the case of new potential bioactive compounds.

In general, it is possible to state that onion is an adequate option in the daily diet to prevent certain pathologies and, in some cases (mainly the extracts), as complementary agents to existing medical treatments. Of particular importance is the fact that, if possible, onion should be consumed as a fresh vegetable or minimally cooked to preserve the most thermolabile bioactive compounds.

On the other hand, during the last years, there has been a clear trend toward the study of bioactivity of onion peel and by-products in order to search new benefits or, at least, the same ones than in the bulb from waste products and, thus, contribute to the sustainability during food processing.

Acknowlegdments

This work has been funded by Ministry of Science and Innovation of Spain (project AGL2007-63462) and by Fun-C-Food CSD2007-00063 Consolider-INGENIO 2010. Marta Corzo-Martínez thanks Danone Institute for a grant.

References

Ali, M., Bordia, T. & Mustafa, T. (1999). Effect of raw versus boiled aqueous extract of garlic and onion on platelet aggregation. *Prostaglandins Leukotrienes and Essential Fatty Acids, 60*, 43-47.

Ali, M., Thomson, M. & Afzal, M. (2000). Garlic and onions: their effect on eicosanoid metabolism and its clinical relevance. *Prostaglandins Leukotrienes and Essential Fatty Acids, 62(2),* 55-73.

Anonymous, WHO monographs on selected medicinal plants, Vol I, 1999, 5-15 (World Health Organization, Geneva).

Ariga, T. & Oshiba, S. (1981). Effects of the essential oil components of garlic cloves on rabbit platelet aggregation. *Igakuto Seibutsugaku, 102,* 169-174.

Arung, E. T., Furuta, S., Ishikawa, S., Kusuma, I. W., Shimizu, K. & Kondo, R. (2011). Anti-melanogenesis properties of quercetin- and its derivative-rich extract from Allium cepa. *Food Chemistry, 124*, 1024-1028.

Ashwah, E. T., El-Allawy, R. M., El-Hashimy, F. S. & Ibrahim, M. H. (1981). The hypoglycemic activity of onion extracts "*Allium cepa*" influenced byadrenaline-induced hyperglycemia. *Journal of Drug Research, 13*, 61-68.

Augusti, K. T. (1973). Studies on the effects of a hypoglycemic principle from Allium cepa L. *Indian Journal of Medical Research, 1*, 1066-1071.

Augusti, K.T. & Mathew, P. T. (1974). Lipid lowering effect of allicin (diallyl disulfide oxide) on long-term feeding in normal rats. *Experientia , 30*, 468-470.

Avila, M. A., Velasco, J. A., Cansado, J. & Notario, V. (1994). Quercetin mediates the down-regulation of mutant p53 in the human breast cancer cell line MDA-MB468. *Cancer Research, 54*, 2424-2428.

Babu, P. S. & Srinivasan, K. (1997). Influence of dietary capsaicin and onion on the metabolic abnormalities associated with streptozotocin induced diabetes mellitus. *Molecular and Cellular Biochemistry, 175*, 49-57.

Bagli, E., Stefaniotou, M., Morbidelli, L, Ziche, M., Psillas, K., Murphy, C. & Fotsis, T. (2004). Luteolin inhibits vascular endothelial growth factor-induced angiogenesis; inhibition of endothelial cell survival and proliferation by targeting phosphatidylinositol 3′-kinase activity. *Cancer Research, 64*, 7936-46.

Ban, J. O., Hwang, I. G., Kim, T. M., Hwang, B. Y., Lee, U. S., Jeong, H. –S., *et al*. (2007). Anti-proliferate and pro-apoptotic effects of 2,3-dihydro-3,5-dihydroxy-6-methyl-4H-pyranone through inactivation of NF-κB in human colon cancer cells. *Archives of Pharmaceutical Research, 30(11),* 1455-1463.

Bang, M. –A. & Kim, H. –A. (2010). Dietary supplementation of onion inhibits diethylnitrosamine-induced rat hepatocellular carcinogenesis. *Food Science of Biotechnology, 19(1),* 77-82.

Bang, M. –A., Kim, H. –A. & Cho, Y. –J. (2009). Alterations in the blood glucose, serum lipids and renal oxidative stress in diabetic rats by supplementation of onion (*Allium cepa* Linn). *Nutrition Research and Practice, 3(3),* 242-246.

Becker, R. C. (1999). Thrombosis and the role of the platelet. *The American Journal of Cardiology*, *83*, 3E-6E.

Benkeblia, N. (2004). Antimicrobial activity of essential oil extracts of various onions (*Allium cepa*) and garlic (*Allium sativum*). *LWT-Food Science Technology, 37*, 263-268.

Benkeblia, N. & Shiomi, N. (2006). Hydrolysis kinetic parameters of DP 6, 7, 8 and 9-12 fructooligosaccharides (FOS) of onion bulb tissues. Effect of temperature and storage time. *Journal of Agricultural and Food Chemistry*, *54*, 2587-2592.

Benítez, V., Mollá, E., Martín-Cabrejas, M.A., Aguilera, Y., López-Andréu, F.J., Cools K., Terry, L.A. Esteban, R.M. (2011). Characterization of industrial onion wastes (Allium cepa L.): Dietary Fibre and bioactive compounds. *Plant Foods Human Nutrition, 66*, 48-57.

Bever, B. O. & Zahnd, G. R. (1979). Plant with oral hypoglycemic action. *The Quarterly Journal of Crude Drug Research, 17*, 139-196.

Bianchini, F. & Vainio, H. (2001). Allium vegetables and organo-sulfur compounds: Do they help prevent cancer? *Environmental Health Perspectives, 109*, 893-902.

Block, E., Birringer, M., Jiang, W., Nakahodo, T., Thompson, H. J., Toscano, H., et al. (2001). *Journal Agricultural and Food Chemistry, 49*, 458.

Block, E., Gulati, H., Putman, D., Sha, D., Niannian, Y. & Zhao, S-H. (1997). *Allium* chemistry: synthesis of 1-[alk(en)ylsulfinyl]-propyl alken(en)yl disulfides (cepaenes), antithrombotic flavorants from homogenates of onion (*Allium cepa*). *Journal of Agricultural and Food Chemistry, 45,* 4414-4422.

Bora, K. S. & Sharma, A. (2009). Phytoconstituents and therapeutic potential of *Allium cepa* Linn. - A Review. *Pharmacognosy Review, 3(5),* 170-180.

Bordia, T., Mohammed, N., Thompson, M. & Ali, M. (1996). An evaluation of garlic and onion as antithrombotic agents. *Prostaglandins, leukotrienes, and essential fatty acids, 54*, 183-186.

Borek, C. (1997). Antioxidants and cancer. *Science and Medicine*, *4*, 51-62.

Borjihan, B., Ogita, A., Fujita, K., Doe, M. & Tanaka, T. (2010). The cyclic organo-sulfur compound Zwiebelane A from onion (*Allium cepa*) functions as an enhancer of Polymyxin B in fungal vacuole disruption. *Planta Medica, 76*, 1864-1866.

Brahmachari, H. D. & Augusti, K. T. (1962). Effect of orally effective hypoglycemic agents from plants on alloxan-induced diabetes. *Journal of Pharmacy and Pharmacology, 14*, 617.

Breu, W. & Dorsch, W. (1994). *Allium cepa* L. (onion), chemistry, analysis and pharmacology. In Wagner, H., and Farnsworth, N. R. (Eds.), *Economic and Medicinal Plant Research* (Vol. 6, pp. 115-147). Academic Press, London.

Briggs, W. H., Folts, J. D., Osman, H. E. & Goldman, I. L. (2001). Administration of raw onion inhibits platelet-mediated thrombosis in dogs. *Journal of Nutrition, 131,* 2619-2622.

Brisdelli, F., Coccia, C., Cinque, B., Cifone, M. G. & Bozzi, A. (2007). Induction of apoptosis by quercetin: different response of human chronic myeloid (K562) and acute lymphoblastic (HSB-2) leukemia cells. *Molecular and Cellular Biochemistry, 296*, 137–149.

Byun, S., Lee K. W., Jung, S. K., Lee, E. J., Hwang, M. K., Lim, S. H., et al. (2010). Luteolin inhibits protein kinase Cε and c-Src activities and UVB-induced skin cancer. *Cancer Research, 70*, 2415-2423.

Cardelle-Cobas, A., Costo, R., Corzo, N. & Villamiel, M. (2009a). Fructo-oligosaccharide changes during the storage of dehydrated commercial garlic and onion samples. *International Journal of Food Science and Technology*, *44*, 947-952.

Cardelle-Cobas, A., Soria, A. C., Corzo-Martínez, M. & Villamiel, M. (2009b). A comprehensive survey of garlic functionality. In *Garlic Compsumtion and Health Research.* Pãcurar, M. and Krejci, G. (Ed). Nova Science Publishers, Inc., New York (USA), pp. 1-60.

Causey, J.L., Feirtag, J.M., Gallaher, D.D., Tungland, B.C. & Slavin, J.L. (2000). Effects of dietary inulin on serum lipids, blood glucose and the gastrointestinal environment in hypercholesterolemic men. *Nutrition Research*, *20*, 191-201.

Cavagnaro, P. F., Sance, M. M. & Galmarini, C. R. (2007). Effect of heating on onion (*Allium cepa* L.) antiplatelet activity and pungency sensory perception. *Food Science and Technology, 13(6),* 447-453.

Chadha, M.C. & Sidhus, A.S. (1990). Studies of the storage life of onion under ambient conditions. Proc of National Symposium on Onion and Garlic 2-3 June, 1990, pp 187-195.

Challier, B., Perarnau, J.-M. Viel, J.-F. (1988). Garlic, onion and cereal fibre as protective factors for breast cancer : A French case-control study. *European Journal of Epidemiology, 14,* 737-747.

Chen, C-H., Chou, T-W., Cheng, L-H. & Ho, C-W. (2011). In vitro anti-adenoviral activity of five *Allium* plants. *Journal of the Taiwan Institute of Chemical Engineers, 42*, 228-232.

Chen, Z.-Y., Peng, C., Jiao, R., Wong, Y. M., Yang, N. & Huang, Y. (2009). Anti-hypertensive nutraceuticals and functional foods. *Journal of Agricultural and Food Chemistry, 57,* 4485–4499.

Chen, J. H., Tsai, S. J. & Chen, H. (1999). Welsh onion (Allium fistulosum L.) extracts alter vascular responses in rat aortae. *Journal of Cardiovascular Pharmacology, 33*, 515-520.

Corzo-Martínez, M., Corzo, N. & Villamiel, M. (2007). Biological properties of onion and garlic. *Trends in Food Science and Technology, 18*, 609-625.

Da Silva, E. L., Tsushida, T. & Terao, J. (1998). Inhibition of mammalian 15-lipoxygenase-dependent lipoprotein by quercetin and quercetin monoglucosides. *Archives of Biochemistry and Biophysics, 49*, 313-320.

De Souza, M.M., Oliveira, M.D., da Rocha, M. & Furlong, E.B. (2010). Antifungal activity evaluation in phenolic extracts from onion, rice bran, and Chlorella phyrenoidosa. *Ciencia e Tecnologia de Alimentos, 30*, 680-685.

Day, A. J., Mellon, F., Barron, D., Sarrazin, G., Morgan, M. R. A. & Williamson, G. (2001). Human metabolism of dietary flavonoids: identification of plasma metabolites of quercetin. *Free Radical Research, 35(6),* 941-952.

Dimitrios, B. (2006). Sources of natural phenolic antioxidants. *Trends in Food Science and Technology, 17*, 505-512.

Dorant, E., Van Den Brandt, P. A., Goldbohm, R. A. & Sturmans, F. (1996). Consumption of onions and a reduced risk of stomach carcinoma. *Gastroenterology, 110*, 12-20.

Dorsch, W. (1996). *Allium cepa* L. (onion) Part 2. Chemistry, analysis and pharmacology. *Phytomedicine*, *3*, 391-397.

Dorsch, W. & Wagner, H. (1991). New antiasthmatic drugs from traditional Medicines? *International Archives of Allergy and Applied Immunology, 94*, 262-265.

El-Demerdash, F. M., Yousef, M. I. & Abou El-Naga, N. I. (2005). Biochemical study on the hypoglycemic effects of onion and garlic in alloxan-induced diabetic rats. *Food and Chemical Toxicology, 43*, 57-63.

Edwards, R. L., Lyon, T., Litwin, S. E., Rabovsky, A., Symons, J. D. & Jalili, T. (2007). Quercetin reduces blood pressure in hypertensive subjects. *Journal of Nutrition, 137*, 2405–2411.

Egert, Bosy-Westphal, Seiberl, Kürbitz, Settler, Plachta-Danielzik, et al. (2009). Quercetin reduces systolic blood pressure and plasma oxidised low-density lipoprotein concentrations in overweight subjects with a high-cardiovascular disease risk phenotype: a double-blinded, placebo-controlled cross-over study. *British Journal of Nutrition, 102*, 1065–1074.

El-Bayoumy, K., Chae, Y. H. & Upadhyaya, P. (2006). Chemoprevention of mammary cancer by diallyl selenide, a novel organoselenium compound. *Anticancer Research, 16,* 2911.

Emmanuel, E. C. & James, O. (2011). Comparative effects of aqueous garlic (*Allium sativum*) and onion (*Allium cepa*) extracts on some haematological and lipid indices of rats. *Annual Review and Research in Biology, 1(3)*, 37-44.

Erlund, I., Kosonen, T., Alfthan, G., Mäenpää, J., Perttunen, K., Kenraali, J., et al. (2000). Pharmacokinetics of quercetin from quercetin aglycone and rutin in healthy volunteers. *European Journal of Clinical Pharmacology, 56,* 545-553.

Ernst, M. & Feldheim, W.J. (2000). Fructans in higher plants and in human nutrition. *Angewandte Botanik*, *74*, 5-9.

Farya, D., Goji, I. A., Joseph, P. K. & Augusti, K. T. (1986). Effects of garlic oil on streptozotocin-diabetic rats maintained on normal and high fat diets. *Indian Journal of Biochemistry and Biophysics, 23,* 24-27.

Fenwick, G.R. & Hanley, A.B. (1985). The genus *Allium.* CRC *Critical Reviews in Food Science and Nutrition*, *22*, 199-377.

Finley, J. W. (2005). Proposed criteria for assessing the efficacy of cancer reduction by plant foods enriched in carotenoids, glucosinolates, polyphenols and selenocompounds. *Annals of Botany, 95*, 1075–96.

Fukushima, S., Takada, N., Hori, T., Min, W. & Wanibuchi, H. (1997). Cancer prevention by organo-sulfur compounds from garlic and onion. *Journal of Cellular Biochemistry. Supplement, 27,* 100-105.

Fukushima, S., Takada, N., Wanibuchi, H., Hori, T., Min, W. & Ogawa, M. (2001). Suppression of chemical carcinogenesis by water-soluble organo-sulphur compounds. *Journal of Nutrition, 131*, 1049s-1053s.

Gabler, N., Osrowska, E., Imsic, M., Eagling, D., Jois, M., Tatham, B. & Dunshea, F. R. (2006). Dietary Onion intake as part of a typical high fat diet improves indices of cardiovascular health using the mixed sex pig model. *Plant Foods for Human Nutrition, 61*, 179-85.

Galeone, C., Pelucchi, C., Levi, F., Negri, E., Franceschi, S., Talamini, R., et al. (2006). Onion and garlic use and human cancer. *American Journal of Clinical Nutrition, 84*, 1027-1032.

Galeone, C., Pelucchia, P., Dal Masoa, L., Negria, E., Montellaa, M., Zucchettoa, A.,et al. (2009a). *Allium* vegetables intake and endometrial cancer risk. *Public Health Nutrition, 12*, 1576-1579.

Galeone, C., Pelucchi, C., Talamini, R., Negri, E., Dal Maso, L.,Montella, M., et al. (2007). Onion and garlic intake and the odds of benign prostatic hyperplasia. *Urology, 70*, 672–676.

Galeone, C., Tavani, A., Pelucchi, C., Negri, E. & La Vecchia, C. (2009b). *Allium* vegetable intake and risk of acute myocardial infarction in Italy. *European Journal of Nutrition, 48,* 120-123.

Gao, C. M., Takezaki, T., Ding, J. H., Li, M. S. & Tajima, K. (1999). Protective effect of *Allium* vegetables against both esophageal and stomach cancer: A simultaneous case-referent study of high-epidemic area in Jiangsu province, China. *Gann, 90*, 614-621.

Gebhardt, R., Beck, H. & Wagner, K. G. (1994). Inhibition of cholesterol biosynthesis by allicin and ajoeno in rat hepatocytes and HepG2 cells. *Biochimica et Biophysica Acta, 1213,* 57-62.

Ghannoum, M.A.J. (1988). Studies on the anticandicidal mode of action of *Allium sativum* (garlic). *Journal of General Microbiology*, *134*, 2917-2924.

Gibson, G.R. (1998). Dietary modulation of the human gut microflora using prebiotics. *The British Journal of Nutrition*, *80*, 209-212.

Glasser, G., Graefe, E. U., Struck, F., Veit, M. & Gebhardt, R. (2002). Comparison of antioxidative capacities and inhibitory effects on cholesterol biosynthesis of quercetin and potential metabolites. *Phytomedicine*, *9*, 33-40.

Gökçe, A.F., Kaya, C., Serçe, S. & Özgen, M. (2010). Effect of scale color on the antioxidant capacity of onions. *Scientia Horticulturae, 123*, 431-135.

Gonzalez, C. A., Pera, G., Agudo, A., Bueno-De-Mesquita, H. B., Ceroti, M., Boeing, H., et al. (2006). Fruit and vegetable intake and the risk of stomach and oesophagus adenocarcinoma in the European Prospective Investigation into Cancer and Nutrition (EPIC-EURGAST). *International Journal of Cancer, 118*, 2559-2566.

Goren, A., Goldman, F., Trainin, Z. & Goldman, R. Antiviral composition derived from *Allium cepa* and therapeutic use thereof. *United States Patent.* Patent No.: US 6, 340, 438 B1. Jan. 22, 2002.

Gorinstein, S., Leontowicz, H., Leontowicz, M., Jastrzebski, Z., Najman, K., Tashma, Z., et al. (2010). The influence of raw and processed garlic and onions on plasma classical and non-classical atherosclerosis indices: investigations *in vitro* and *in vivo*. *Phytotherapy Research, 24,* 706–714.

Graefe, E.U., Witting, J., Mueller, S., Riethling, A. K., Uehleke, B., Drewelow, B., et al. (2001). Pharmacokinetics and bioavailability of quercetin glycosides in human. *Journal of Clinical Pharmacology, 41,* 492-499.

Griffiths, G., Trueman, L., Crowther, T., Thomas, B. & Smith, B. (2002). Onions – A global benefit to health. *Phytotherapy Research*, *16*, 603-615.

Gülsen, A., Makris, D. P. & Kefalas, P. (2007). Biomimetic oxidation of quercetin: Isolation of a naturally occurring quercetin heterodimer and evaluation of its in vitro antioxidant properties. *Food Research International, 40*, 7-14.

Gupta, N. & Porter, T. D. (2001). Garlic and garlic-derived compounds inhibit human squalene monooxygenase. *The journal of Nutrition*, *131*, 1662-1667.

Guyonnet, D., Belloir, C., Suschetet, M., Siess, M. –H. & Le Bon, A. –M. (2001). Antimutagenic activity of organo-sulfur compounds from *Allium* is associated with phase II enzyme induction. *Mutation Research: Genetic Toxicology and Environmental Mutagenesis, 495*, 35-145.

Gutteridge, J.M.C. (1993). Free radicals in desease processes: a complication of cause and consequence. *Free Radical Research Communications*, *19*, 141-158.

Ho, S.-C., Tamg, Y.-L., Lin, S.-M. & Liew, Y.-F. (2010). Evaluation of peroxynitrite-scavenging capacities of several commonly used fresh spices. *Food Chemistry, 119,* 1102–1107.

Hollman, P.C.H., Devries, J.H.M., Vanleeuwen, S. D., Mengelers, M. J. B. & Katan, M. B. (1995). Absorption of dietary quercetin glycosides and quercetin in healthy ileostomy volunteers. *The American Journal of Clinical Nutrition, 62,* 1276-1282.

Hsing, A. W., Chokkalingam, A. P., Gao, Y. T., Madigan, M. P., Deng, J., Gridley, G., et al. (2002). *Allium* vegetables and risk of prostate cancer: A population-based study. *J. Natl. Cancer Inst. 94*, 1648-1651.

Hu, J., La Vecchia, C., Negri, E., Chatenoud, L., Bosetti, C., Jia, X., et al. (1999). Diet and brain cancer in adults. A case-control study in northeast China. *International Journal of Cancer, 81,* 20-23.

Huang, Y. T., Hwang, J. J., Lee, P. P., et al. (1999). Effects of luteolin and quercetin, inhibitors of tyrosine kinase, on cell growth and metastasis-associated properties in A431 cells overexpressing epidermal growth factor receptor. *British Journal of Pharmacology, 128,* 999–1010.

Hubbard, G. P., Wolffram, S., de Vos, R., Bovy, A., Gibbins, J. M. & Lovegrove, J. A. (2006). Ingestion of onion soup high in quercetin inhibits platelet aggregation and essential components of the collagen-stimulated platelet activation pathway in man: a pilot study. *British Journal of Nutrition, 96*, 482-488.

Hurst, R., Armah, C. N., Dainty, J. R., Hart, D. J., Goldson, A. J., Broadley, M. R., et al. (2010). Establishing optimal selenium status: results of a randomized, double-blind, placebo-controlled trial. *The American Journal of Clinical Nutrition, 91*, 923–31.

Ip, C., Lisk, D. J. & Stoewsand, G. S. (1992). Mammary cancer prevention by regular garlic and selenium-enriched gralic. *Nutrition Cancer, 17,* 279-285.

Irkin, R. & Korukluoglu, M. (2007). Control of *Aspergillus niger* with garlic, onion and leek extracts. *African Journal of Biotechnology,* 6, 384-387.

Irkin, R. & Korukluoglu, M. (2009). Control of some filamentous fungi and yeasts by dehydrated Allium extracts. *Journal of Consumer Protection and Food Safety, 4*, 3-6.

Jan, A.T., Kamli, M.R., Murtaza, I., Singh, J.B., Ali, A. & Haq, Q.M.R. (2010). Dietary Flavonoid quercetin and associated health benefits-An overview. *Food Reviews International, 26*, 302-3017.

Jain, R. C. & Vyas, C. R. (1974). Hypoglycemic action of onion on rabbits. *British Medical Journal, 2*, 730.

Jain, R. C. & Vyas, C. R. (1977). Onion and Garlic in atherosclerotic heart disease. *Medikon, 6,* 12-14.

Jang, J. R. & Lim, S. –Y. (2009). Effects of onion flesh and peel on chemical components, antioxidant and anticancer activities. *Korean Journal of Life Science, 19(11),* 1598-1604.

Janssen, K., Mensink, R. P., Cox, F. J. J., Harryvan, J. L., Hovenior, R. & Hollman, P. C. H. (1998). Effects of the flavonoids quercetin and apigenin on hemostasis in healthy volunteers: results from an in vitro and a dietary supplement study. *The American Journal of Clinical Nutrition, 67*, 255-262.

Jeong, C. H., Heo, H. J., Choi, S. G. & Shim, K. H. (2009). Antioxidant and anticancer properties of methanolic extracts from different parts of white, yellow and red onion. *Food Science and Biotechnology, 18(1),* 108-112.

Jiemei, Z., Jianjun, H., Jie, G., Benhong, H. & Hao, H. Effects of fistular onion stalk extract on the level of NO and expression of endothelial NO synthase (eNOS) in human umbilical vein endothelium cells. *African Journal of Biotechnology, 10(13)*, 2536-2540.

Jung, Y. S., Kim, M. H., Lee, S. H., Baik, E. J., Park, S.W. & Moon, C. H. (2002). Antithrombotic effect of onion in streptozotocin-induced diabetic rat. *Prostaglandins, Leukotrienes and Essential Fatty Acids, 66*, 453-458.

Jung, J. Y., Lim, Y., Moon, M. S., Kim, J. Y. & Kwon, O. (2011). Onion peel extracts ameliorate hyperglycemia and insulin resistance in high fat diet/streptozotocin-induced diabetic rats. *Nutrition and Metabolism, 8*, 18.

Kaiser, P., Youssouf, S., Tasduq, S.A., Singh, S., Sharma, S.C., Singh, G.D., Gupta, V.K., Gupta, B.D., Johri, R.K. (2009). Antiallergic effects of herbal product from *Allium cepa* (Bulb). *Journal of Medicinal Science, 12*, 374-382.

Kaneko, T. & Baba, N. (1999). Protective effect of flavonoids on endothelial cells against linoleic acid hydroperoxide-induced toxicity. *Bioscience, Biotechnology and Biochemistry, 63*, 323-328.

Karawya, M. S., Abde, S. M., El-Olemy, M. M. & Farrag, N. M. (1984). Diphenylamine, an antihyperglycemic agent from onion and tea. *Journal of Natural Products, 47*, 775-780.

Khaki, A., Fathiazad, F., Nouri, M., Khaki, A.A., Khamenehi, H.J. & Hamadeh, M. (2009). Evaluation of androgenic activity of allium cepa on spermatogenesis in the rat. *Folia Morphol 68*, 45-51.

Khiari, Z., Makris, D.P. & Kefalas, P. (2009). An investigation on the recovery of antioxidant phenolics from onion solid wastes employing water/ethanol-based solvent systems. *Food and Bioprocess Technology, 2*, 337-343.

Kim, J. Y. & Kwon, O. (2009). Garlic intake and cancer risk: An analysis using the Food and Drug Administration's evidence-based review system for the scientific evaluation of health claims. *The American Journal of Clinical Nutrition, 89*, 257-264.

King, M. W. & Marchesisni, S. (2007). Cholesterol and Bile Metabolism. Infornmation online, available at; http://med.unibs.it/~marchesi/ cholesthtml.

Kotrebai, M., Tyson, J. F., Uden, P. C., Birringer, M. & Block, E. (2000). Selenium speciation in enriched and natural samples by HPLC-ICP-MS and HPLC-ESI-MS with perfluorinated carboxylic acid ion-pairing agents. *Analyst, 125,* 71-78.

Kuhlmann, M. K., Burkhardt, G., Horsch, E., Wagner, M. & Kohler, H. (1998). Inhibition of oxidant-induced lipid peroxidation in cultured renal tubular epithelial cells (LLC-PK1) by quercetin. *Free Radical Research, 29,* 451-460.

Kumari, K. & Augusti, K. T. (2007). Lipid lowering effect of S-methyl cysteine sulfoxide from *Allium cepa* Linn in high cholesterol diet fed rats. *Journal of Ethnopharmacology, 109,* 367-371.

Kumari, K. & Mathew, B. M. (1995). Antidiabetic and hypolipidemic effects of S-Mehyl cysteine sulphoxide isolated from *Allium cepa* L. *Indian Journal of Biochemistry and Biophysics, 32,* 49-54.

Lancaster, J. E. & Shaw, M. L. (1989). G-Glutayl peptides in the biosynthesis of S-alk(en)yl-L-cysteine sulfoxides (flavor precursors) in *Allium. Phytochemistry*, *28*, 455-460.

Lanzotti, V. (2006). The analysis of onion and garlic. *Journal of Chromatography A, 1112,* 3-22.

Lata, S., Saxena, K. K., Bhasin, V., Saxena, R. S., Kumar, A. & Srivastava, N. K. (1991). Beneficial effects *Alliums sativa, Alliums cape*, and *Commiphora mukul* on experimental hyperlipidemia and atherosclerosis: A comparative evaluation. *Journal of Postgraduate Medicine, 37*, 132-135.

Lautraite, S., Musonda, A. C., Doehmer, J., Edwards, G. O. & Chipman, J. K. (2002). Flavonoids inhibit genetic toxicity produced by carcinogens in cells expressing CYP1A2 and CYP1A1. *Mutagenesis, 17*, 45-53.

Lean, M., Noroozi, M., Kelly, I., Burns, J., Talwar, D. & Satter, N. (1999). Dietary flavonoids protect diabetic human lymphocytes against oxidant damage to DNA. *Diabetes, 48,* 176-178.

Lee, S. K., Hwang, J. Y., Kang, M. J., Kim, Y. M., Jung, S. H., Lee, J. H. & Kim, J. I. (2008). Hypoglycemic effect of onion skin extract in animal models of diabetes mellitus. *Food Science and Biotechnology, 17,* 130-134.

Lee, Y.R., Hwang, I.G., Woo, K.S., Kim, D.J., Hong, J.T. & Jeong, H.S. (2007). Antioxidative activities of the ethyl acetate fraction from heated onion (*Allium cepa*). *Food Science and Biotechnology, 16*, 1041-1045.

Lee, K-A., Kim, K-T., Nah, S-Y., Chung, M-S., Cho, S. & Paik, H-D. (2011a). Antimicrobial and antioxidative effects of onion peel extracted by the subcritical water. *Food Science and Biotechnology*, *20*, 543-548.

Lee, H. –J., Lee, K. –H., Park, E. & Chung, H. K. (2010). Effect of onion extracts on serum cholesterol in borderline hypercholesterolemic participants. Journal of the *Korean Society of Food Science and Nutrition, 39(12),* 1783-1789.

Lee, K. –H., Park, E., Lee, H. –J., Kim, M. –O., Cha, Y. –J., Kim, J. –M., et al. (2011b). Effects of daily quercetin-rich supplementation on cardiometabolic risks in male smokers. *Nutrition Research and Practice, 5(1),* 28-33.

Lee, W. J., Wu, L. F., Chen, W. K., Wang, C. J. & Tseng, T. H. (2006). Inhibitory effect of luteolin on hepatocyte growth factor/scatter factor-induced HepG2 cell invasion involving both MAPK/ERKs and PI3K-Akt pathways. *Chemico-Biological Interactions, 160,* 123-33.

Le Marchand, L. (2002). Cancer preventive effects of flavonoids- a review. *Biomedicine and Pharmacotherapy*, *56*(6), 296-301.

Le Marchand, L., Murphy, S. P., Hankin, J. H., Wilkens, L. R. & Kolonel, L. N. (2000). Intake of flavonoids and lung cancer. *Journal of the National Cancer Institute, 92,* 154-160.

Levi, F., La Vecchia, C., Gulie, C. & Negri, E. (1993). Dietary factors and breast cancer risk in Vaud, Switzerland. *Nutrition and Cancer, 19*, 327-335.

Lotito, S. B. & Frei, B. (2006). Consumption of flavonoid-rich foods and increased plasma antioxidant capacity in humans: cause, consequence, or epiphenomenon? *Free Radical Biology and Medicine, 41*, 1727-1746.

Ly, T.N., Hazama, C., Shimoyamada, M., Ando, H., Kato, K. & Yamauchi, R. (2005). Antioxidative compounds from the outer scales of onion. *Journal of Agricultural and Food Chemistry, 53,* 8183-8189.

Makheja, A. & Baily, J. Antiplatelet constituents of garlic and onion. (1990). *Agents and Actions, 29*, 360-363.

Makheja, A. N., Vanderhoek, J. Y. & Bailey, J. M. (1979). Inhibition of platelet aggregation and thromboxane synthesis by onion and garlic. *Lancet, 1*, 781.

Malaveille, C., Hautefeuille, A., Pignatelli, B., Talaska, G., Vineis, P. & Bartsch, H. (1996). Dietary phenolics as anti-mutagens and inhibitory of tobacco related DNA adduction in the urothelium of smokers. *Carcinogenesis, 17*, 2193-2200.

Manju, V., Balasubramaniyan, V. & Nalini, N. (2005). Rat colonic lipid peroxidation and antioxidant status: the effects of dietary luteolin on 1,2-dimethylhydrazine challenge. *Cellulr and Molecular Biology Letters, 10*, 535-51.

Matsuura, H. (1997). Phytochemistry of garlic horticultural and processing procedures. In Lachance, P. A. (Ed), *Neutraceuticals: designer foods III. Garlic, soy and licorice* (pp. 55-59). Trumbull, C. T: Food and Nutrition Press.

Matsuura, H. (2001). Saponins in garlic as modifiers of the risk of cardiovascular disease. *The Journal of Nutrition*, *131*,1000S-1005S.

Mayer, B., Kalus, U., Grigorov, A., Pindur, G., Jung, F., Radtke, H., *et al.* (2001). Effects of an onion-olive oil maceration product containing essential ingredients of the Mediterranean diet on blood pressure and blood fluidity. *ArzneimittelForschung, 51(2)*, 104-111.

Millen, A. E., Subar, A. F., Graubard, B. I., Peters, U., Hayes, R. B., Weissfeld, J. L., et al. for the PLCO Cancer Screening Project Team. (2007). Fruit and vegetable intake and prevalence of colorectal adenoma in a cancer screening trial. *American Journal of Clinical Nutrition, 86*, 1754-1764.

Miodini, P., Gioravanti, L., Di Fronzo, G. & Cappelletti, V. (1999). The two phyto-oestogens genistein and quercetin exert different effects on oestrogen receptor function. *British Journal of Cancer, 80*, 1150-1155.

Miron, T., Rabinkov, E., Peleg, T., Rosenthal, D., Mirelman, M. & Wilchek, M. (2004). Allylmercaptocaptopril: a new antihypertensive drug. *American Journal of Hypertension*, *17*, 71-73.

Moon, C. H., Jung, Y. S., Kim, M. H., Lee, S. H., Baik, E. J. & Park, S. W. (2000). Mechanism for antiplatelet effect of onion: araechidonic acid release inhibition, thromboxane A(2) synthase inhibition and TXA (2)/PGH (2) receptor blockade. *Prostaglandins Leukotrienes and Essential Fatty Acids*, *62*, 277-283.

Moreno, F.J., Corzo-Martínez, M., del Castillo, M.D. & Villamiel, M. (2006). Changes in antioxidant activity of dehydrated onion and garlic during storage. *Food Research International*, *39*, 891-897.

Musonda, C. A. & Chipman, J. K. (1998). Quercetin inhibits hydrogen peroxide (H_2O_2)-induced NF-kappaB DNA binding activity and DNA damage in HepG2 cells. *Carcinogenesis, 19*, 1583-1589.

Muto, S., Fujita, K.-I. & Yamazaki, T. (2001). Inhibition by green tea catechins of metabolic activation of procarcinogens by human cytochrome P450. *Mutation Research*, 479, 197-206.

Mutoh, M., Takashi, M., Fukuda, K., Komatsu, H., Enya, T., Masushima-Hibiya, Y., et al. (2000). Suppression by flavonoids of cyclooxygenase-2 promoter-dependent transcriptional activity in colon cancer cells: structure-activity relationship. *Japanese Journal of Cancer Research, 91*, 686-691.

Naseri, M. K., Arabian, M., Badavi, M. & Ahangarpour, A. (2008a). Vasorelaxant and hypotensive effects of Allium cepa peel hydroalcoholic extract in rat. *Pakistan Journal of Biological Sciences, 11*, 1569–1575.

Naseri, M. K., Yahyavi, H. & Arabian, M. (2008b). Antispasmodic activity of onion (*Allium cepa* L.) peel extract on rat ileum. *Iranian Journal of Pharmaceutical Research,* 7, 155-159.

Ostrowska, E., Gabler, N. K., Sterling, S. J., Tatham, B. G., Jones, R. B., Eagling, D. R. et al. (2004). Consumption of brown onions (*Allium cepa* var. cavalier and var. density) moderately modulates blood lipids, haematological and haemostatic variables in healthy pigs. *British Journal of Nutrition, 91*, 211-218.

Park, J. B. (2011). Effects of typheramide and alfrutamide found in *Allium* species on cyclooxygenases and lipoxygenases. *Journal of Medical Food, 14(3)*, 226-231.

Park, J., Kim, L. & Kim, M. K. (2007). Onion flesh and onion peel enhance antioxidant ssstatus in aged rats. *Journal of Nutrition Science and Vitaminology, 53*, 21-29.

Park, S.Y. & Chin, K.B. (2010). Effects of onion on physicochemical properties, lipid oxidation and microbial growth of fresh pork patties. *International Journal of Food Science and Technology, 45*,1153-1160.

Park, S.Y., Yoo, S.S., Shim, J.H. & Chin, K.B. (2008). Physicochemical properties, and antioxidant and antimicrobial effects of garlic and onion powder in fresh pork belly and loin during storage. *Journal of Food Science*, *73*, C577-C583.

Perchellet, J. P., Perchellet, E. M. & Bellman, S. (1990). Inhibition of DMBA-induced mouse skin tumorigenesis by garlic oil and inhibition of two tumor-promotion stages by garlic and onion oil. *Nutrition and Cancer*, *14*, 183-193.

Pérez-Gregorio, M.R., Regueiro, J., González-Barreiro, C., Rial-Otero, R. & Simal-Gándara J. (2011) Changes in antioxidant flavonoids during freeze-drying of red onions and subsequent storage. *Food Control, 22,* 1108-1113.

Phillippe, B., Cammue, B.P.A., Thevissen, K., Hendriks, M., Eggermont, K., Goderis, I. J., et al. (1995). A potent antimicrobial protein from onion seeds showing sequence homology to plant lipid transfer proteins. *Journal of Plant Physiology*, *109*, 445-455.

Platel, K. & Srinivasan, K. (2001). Studies on the influence of dietary spices on food transit time in experimental rats. *Nutrition Research*, *21*, 1309-1314.

Platt, K. L., Edenharderb, R., Aderholda, S., Muckelc, E. & Glattc, H. (2010). Fruits and vegetables protect against the genotoxicity of heterocyclic aromatic amines activated by human xenobiotic-metabolizing enzymes expressed in immortal mammalian cells. *Mutation Research, 703,* 90–98.

Pradhan, S. J., Mishra, R., Sharma, P. Kundu, G. C. (2010). Quercetin and sulforaphane in combination suppress the progression of melanoma through the down-regulation of matrix metalloproteinase-9. *Experimental and Therapeutic Medicine, 1*, 915-920.

Pszczola, D.E. (2002). Antimicrobials: setting up additional hurdles to ensure food safety. *Food and Technology, 56,* 99-107.

Rahman, K., Allison, G. L. & Lowe, G. M. (2006). Mechanisms of inhibition of platelet aggregation by aged garlic extract and its constituents. *The Journal of Nutrition, 136,* 782S-788S.

Ramos, F.A., Takaishi, Y., Shirotori, M., Kawaguchi, Y., Tsuchiya, K., Shibata, H., et al. (2006). Antibacterial and antioxidant activities of quercetin oxidation products from yellow onion (*Allium cepa*) skin. *Journal of Agricultural and Food Chemistry, 54,* 3551-3557.

Raso, G. M., Meli, R., Di Carlo, G., Pacilio, M. & Di Carlo, R. (2001). Inhibition of inducible nitric oxide synthase and cyclooxygenase-2 expression by flavonoids in macrophage J774A.1. *Life Sciences, 68*, 921-931.

Reddy, B. S., Rao, C. V., Rivenson, A. & Kelloff, G. (1993). Chemoprevention of colon carcinogenesis by organo-sulfur compounds. *Cancer Research, 53*, 3493-3498.

Richardson, S.J. (1993). Free radicals in the genesis of Alzheimer's disease. *Annals of the New York Academy of Sciences*, *695*, 73-76.

Roldán, E., Sánchez-Moreno, C., de Ancos, B., Cano, P. & Cano, M.P. (2008) Characterization of onion (*Allium cepa* L.) by products as food ingredients with antioxidant and antibrowning properties. *Food Chemistry, 108,* 907-916.

Roldán-Marín, M. E. (2009). Biological activity and nutritional properties of processed onion products. PhD Thesis.

Roldán-Marín, E., Jensen, R. I., Krath, B. N., Kristensen, M., Poulsen, M., Cano, M. P., et al. (2010). An onion byproduct affects plasma lipids in healthy rats. *Journal of Agricultural and Food Chemistry, 58,* 5308-5314.

Roldán-Marín, E., Krath, B.N., Poulsen, M., Binderup, M-L., Nielsen, T-H., Hansen, M., Barri, T., Langkilde, S., Cano, M.P., Sánchez-Moreno, C. & Dragsted, L.O. (2009). *British Journal of Nutrition, 102*, 1574-1582.

Rose, P., Whiteman, M., Moore, P. K. & Zhu, Y. Z. (2005). Bioactive S-alk(en)yl cysteine sulfoxide metabolites in the genus Allium: the chemistry of potential therapeutic agents. *Natural Product Reports*, *22*, 351-368.

Rose, P., Widder, S., Looft, J., Pickenhagen, W., Ong, C. N. & Whiteman, M. (2003). Inhibition of peroxynitrite-mediated cellular toxicity, tyrosine nitration, and R1-antiproteinase inactivation by 3-mercapto-2-methylpentan-1-ol, a novel compound isolated from *Allium cepa. Biochemical and Biophysical Research Communications, 302*, 397–402.

Roy, M.K., Takenaka, M. & Isobe, S. (2007.) Anti-radical activity and reduced pro-oxidant.Activity in water-soluble fraction of selected Allium vegetables. *Journal of the Science of Food and Agriculture, 87,* 2259-2265.

Sainani, G. S., Desai, D. B. & Sainine, P. (1978). Onion in prevention of atherosclerosis. *Journal of Indian Medical Association, 71,* 109.

Sakai, Y., Murakami, T. & Yamamoto, Y. (2003). Antihypertensive effects of onion on NO synthase inhibitor-induced hypertensive rats and spontaneously hypertensive rats. *Biosci., Biotechnol., Biochem., 67*, 1305–1311.

Saleheen, D., Ali, S. A. & Yasinzai, M. M. (2004). Antileishmanial activity of aqueous onion extract in vitro. *Fitoterapia*, *75*, 9-13.

Sanchez, M., Lodi, F., Vera, R., Villar, I. C., Cogolludo, A., Jimenez, R., et al. (2007). Quercetin and isorhamnetin prevent endothelial dysfunction, superoxide production, and overexpression of p47(phox) induced by angiotensin II in rat aorta. *The Journal of Nutrition, 137,* 910-915.

Sankaranarayanan, R., Varghese, C., Duffy, S. W., Padmakumary, G., Day, N. E. & Nair, M. K. (1994). A case-control study of diet and lung-cancer in Kerala, South-India. *International Journal of Cancer, 58*, 644-649.

Santas, J., Almajano, M.P. and Carbo, R. (2010). Antimicrobial and antioxidant activity of crude onion (*Allium cepa*, L.) extracts. *International Journal of Food Science and Technology, 45,* 403-409.

Saxena, A., Tripathi, R.M. & Singh, R.P. (2010). Biological synthesis of silver nanoparticles by using onion (*Allium cepa*) extract and their antibacterial activity. *Digest Journal of Nanomaterilas and Biostructures, 5*, 427-432.

Sellappan, S. & Akoh, C. C. (2002). Flavonoids and antioxidant capacity of Georgia-grown Vidalia onions. *Journal of Agricultural and Food Chemistry, 50,* 5338-5342.

Sharma, K. K., Chowdhury, N. K. & Sharma, A. L. (1975). Studies on hypocholesterolaemic activity of onion. II. Effect on serum cholesterol in rabbits maintained on high cholesterol diet. *Indian Journal of Nutrition and Dietetics, 12*, 388-391.

Sharma, A. D., Kainth, S. & Gill, P. K. (2006). Inulinase production using garlic (*Allium sativum*) powder as a potential substrate in *Streptomyces* sp. *Journal of Food Engineering*, *77*, 486-491.

Sharma, K. K. & Sharma, S. P. (1976). Effect of onion on blood cholesterol, fibrinogen and fibrinolytic activity in normal subjects. *Indian Journal of Pharmacology, 8*, 231-233 (1976).

Sharma, K. K. & Sharma, S. P. (1979). Effect of onion and garlic on serum cholesterol on normal subjects. *Mediscope, 22*, 134-136.

Sheela, C. G. & Augusti, K. T. (1995). Antiperoxide effects of S-allyl cysteine sulphoxide isolated from Allium sativum Linn and gugulipid in cholesterol diet fed rats. *Indian Journal of Experimental Biology, 33*, 337-341.

Shen, F., Herenyiova, M. & Weber, G. (1999). Synergistic down-regulation of signal transduction and cytotoxicity by tiazofurin and quercetin in human ovarian carcinoma cells. *Life Sciences, 64,* 1869-1876.

Shin, S- C., Choi, J- S. & Li, X. (2006). Enhanced bioavailability of tamoxifen after oral administration of tamoxifen with quercetin in rats. *International Journal of Pharmaceutic, 313,* 144-149.

Singh, B.N., Singh, B.R., Singh, R.L., Prakash, D., Singh, D.P., Sarma, B.K., et al. (2009). Polyphenolics from various extracts/fractions of red onion (*Allium cepa*) peel with potent antioxidant and antimutagenic activities. *Food and Chemical Toxicology, 47,* 1161-1167.

Škerget M, Majhenič L, Bezjak M. & Knez Z (2009). Antioxidant, radical scavenging and antimicrobial activities of red onion (*Allium cepa* L) skin and edible part extracts. *Chemical and Biochemical Engineering Quaterly, 23*, 435-444.

Sobenin, I. A., Tertov, V. V. & Orekhov, A. N. (1994). Characterization of chemical composition of native and modified low dendity lipoproteins occurring in the blood of diabetic patients. *International Angiology, 13,* 78-83.

Srinivasan, K. (2004a). Plant foods in the management of diabetes mellitus: Spices as potential antidiabetic agents. *International Journal of Food Science of Nutrition, 56(6)*, 399-414.

Srinivasan, K. (2005). Plants food in the management of diabetes mellitus: spices as potencial antidiabetic agents. *International Journal of Food Science and Nutrition*, *56,* 399-414.

Srinivasan, K. & Sambaiah, K. (1991). The effect of spices on cholesterol 7 alpha-hydroxylase activity and on serum and hepatic cholesterol levels in the rat. *International Journal of Vitamin and Nutrition Research, 61,* 364-369.

Srinivasan, K., Sambaiah, K. & Chandrasekhara, N. (2004b). Spices as beneficial hypolipidemic food adjuncts: A Review. *Food Reviews International, 20,* 187-220.

Srivastava, K. C. (1984). Aqueous extracts of onion, garlic and ginger inhibit platelet aggregation and alter arachidonic acid metabolism. *Biomedica Biochimica Acta, 43,* 335-346.

Steerenberg, P. A., Garssen, J., Dortant, P., Hollman, P. C., Alink, G. M. & Dekker, M. (1998). Protection of UV-induced suppression of skin contact hypersensitivity: A common feature of flavonoids after oral administration? *Photochemistry and Photobiology, 67*, 456-461.

Steinmetz, K. A., Kushi, L. H., Bostick, R. M., Folsom, A. R. & Potter, J. D. (1994). Vegetables, fruit and colon cancer in the lowa woman's health study. *American Journal of Epidemiology, 139*, 1-15.

Taché, S., Ladam, A. & Corpet, D. E. (2007). Chemoprevention of aberrant crypt foci in the colon of rats by dietary onion. *European Journal of Cancer, 43,* 454-458.

Takada, N., Yano, Y., Wanibuchi, H., Otani, S. & Fukushima, S. (1997). S-methylcysteine and cysteine are inhibitors of induction of glutathione S-transferasa placental from-positive foci during initiation and promotion phase of rat hepatocarcogenesis. *Japanese Journal of Cancer Research, 88,* 435-442.

Tansey, M.R. & Appleton, J.A. (1975). Inhibition of fungal growth by garlic extract. *Mycologia*, *67*, 409-413.

Taran, M., Rezaeian, M. & Izaddoost, M. (2006). *In vitro* antitrichomonas activity of *Allium hirtifolium* (Persian Shallot) in comparison with metronidazole. *Iranian Journal of Public Health, 35*, 92-94.

Terao, J., Kawai, Y. & Murcita, K. (2008). Vegetable flavonoids and cardiovascular disease. *Asia Pacific Journal of Clinical Nutrition, 17,* 291-293.

Tsao, S.M. & Yin, M.C. (2001). In-vitro antimicrobial activity of four diallyl sulphides occurring naturally in garlic and Chinese leek oils. *Journal of Medical Microbiology*, *50*, 646-649.

Tsuda, H., Ohshima, Y., Nomoto, H., Fujita, K.-I., Matsuda, E., Iigo, M., et al. (2004). Cancer prevention by natural compounds. *Drug Metabolism and Pharmacokinetics, 19(4)*, 245-263.

Ueda, H., Yamazaki, C. & Yamazaki, M. (2002). Luteolin as an anti-inflammatory and anti-allergic constituent of Perilla frutescens. *Biological and Pharmaceutical Bulletin, 25*, 1197–202.

Ueda, H., Yamazaki, C. & Yamazaki, M. (2003). Inhibitory effect of perilla leaf extract and luteolin on mouse skin tumor promotion. *Biological and Pharmaceutical Bulletin, 26,* 560–563.

Urios, P., Grigorova-Borsos, A-M. & Sternberg, M. (2007). Flavonoids inhibit the formation of the cross-linking AGE pentosidine in collagen incubated with glucose, according to their structure. *European Journal of Nutrition, 46,* 139-146.

Van Damme, E.J.M., Willems, P., Torrekans, S., Van Leuven, F. & Peumans, W.J. (1993). Garlic (*Allium sativum*) chitinase characterization and molecular cloning. *Physiologia Plantarum, 187*, 177-186.

Vanderhoek, J. Y., Makheja, A. N. & Bailey, J. M. (1980). Inhibition of fatty acid oxygenases by onion and garlic oils. Evidence for the mechanism by which these oils inhibit platelet aggregation. *Biochemical Pharmacology, 29,* 3169-3173.

Vatsala, T. M., Singh, M. & Murugesan, R. G. (1980). Effects of onion in induced atherosclerosis in rabbits: In Reduction of arterial lesions and lipid levels. *Artery, 7,* 519-530.

Vijayababu, M. R., Arunkumar, A., Kanagaraj, P., Venkataraman, P., Krishnamoorthy, G. & Arunakaran, J. (2006). Quercetin downregulates matrix metalloproteinases 2 and 9 proteins expression in prostate cancer cells (PC-3). *Molecular and Cellular Biochemistry, 287*, 109-116.

Viry, E., Anwar, A., Kirsch, G., Jacob, C., Diederich, M. & Bagrel, D. (2011). Antiproliferative effect of natural tetrasulfides in human breast cancer cells is mediated through the inhibition of the cell division cycle 25 phosphatases. *International Journal of Oncology, 38(4),* 1103-1111.

Wagner, H., Dorsch, W., Bayer, T., Breu, W. & Willer F. (1990). Antiasthmatic effects of onions: inhibition of 5-lipoxygenase and cyclooxygenase *in vitro* by thiosulfinates and "cepaenes". *Prostaglandins Leukotrienes and Essential Fatty Acids*, *39*, 59-62.

Wang, H.X. & Ng, T.B. (2004). Isolation of allicepin, a novel antifungal peptide from onion (*Allium cepa*) bulbs. *Journal of Peptide Science*, *10*, 173-177.

Wargovich, M.J., Woods, C., Eng, V.W.S., Stephens, C. & Gray, K. (1988). Chemoprevention of N-nitroso, methylbenzylamine-induced esophageal cancer in rats by naturally occurring thioether, diallyl sulfide. *Cancer Research*, *48*, 6872-6875.

Wensing, M., Ludt, S., Campbell, S., van Lieshout, J., Volbracht, E., Grol, R., on behalf of the EPACPG. (2009). European practice Aassessment of cardiovascular risk management (EPA Cardio): protocol of an international observational study in primary care. *Implementation Science, 4,* 3-10.

WHO (World Health Organization) Cardiovascular disease (http://www.who.int/cardiovascular_diseases/en/. Accesed: 24 October, 2009).

WHO monographs on selected medicinal plants, Vol I, (World Health Organization, Geneva, 1999) 5-15.

Wilcox, B. F., Joseph, P. K. & Augusti, K. T. (1984). Effects of allylpropyl disulphide isolated from *Allium cepa* Linn on high-fat fed rats. *Indian Journal of Biochemistry and Biophysics, 21*, 214-216.

Winocour, P. D. (1994). Platelets, vascular disease, and diabetes mellitus. *Canadian Journal of Physiology and Pharmacology*, *72*, 295-303.

Woo, K.S., Hwang, I.G., Kim, T.M., Kim, D.J., Hong, A.T. & Jeong, H.S. (2007). Changes in the antioxidant activity of onion (*Allium cepa*) extracts with heat treatment. *Food Science and Biotechnology, 16*, 828-831.

Wu, C. P., Calcagno, A. M., Hladky, S. B., Ambudkar, S. V. Barrand, M. A. (2005). Modulatory effects of plant phenols on human multidrug-resistance proteins 1, 4 and 5 (ABCC1, 4 and 5). *FEBS Journal, 272(18),* 4725-4740.

Wu, Y., He, Y. & Ge, X. (2011). Functional characterization of the recombinant antimicrobial peptide Trx-Ace-AMP1 and its application on the control of tomato early blight disease. *Applied Microbiology and Biotechnology, 90*, 1303-1310.

Xiao, H. & Parkin, K. L. (2007). Isolation and identification of potential cancer chemopreventive agents from methanolic extracts of green onion (*Allium cepa*). *Phytochemistry*, 68, 1059–1067.

Yamamoto, Y., Aoyama, S., Hamaguchi, N. Rhi, G. S. (2005). Antioxidative and antihypertensive effects of Welsh onion on rats fed with a high-fat high-sucrose diet. *Bioscience, Biotechnology, and Biochemistry, 69,* 1311–1317.

Yang, J., Meyers, K.J., Van der Heide, J. & Liu, R.H. (2004). Varietal differences in phenolic content and antioxidant and anti proliferative activities of onions. *Journal of Agricultural and Food Chemistry*, *52*, 6787-6793.

You, W. C., Zhang, L., Gail, M. H., Ma, J. L., Chang, Y. S., Blot, W. J., et al. (1998). *International Journal of Epidemiology, 27(6)*, 941-944.

Zohri, A.N., Abdel-Gawad, K. & Saber, S. (1995). Antibacterial, antidermatophytic and antioxigenic activities of onion (*Allium cepa L.*) oil. *Microbiological Research*, *150*, 167-172.

Zouhir, A-M., Kheadr, E., Tahiri, I., Ben Hamida, J. & Fliss, I. (2008). Combination with plnat extract improves the inhibitory action of divergicin M35 agaisnt *Listeria monocytogenes. Journal of Food Quality, 31*, 13-33.

In: Onion Consumption and Health
Editors: C. B. Aguirre et al.
ISBN: 978-1-62100-836-1

Chapter 2

ONION MAJOR COMPOUNDS (FLAVONOIDS, ORGANOSULFURS) AND HIGHLY EXPRESSED GLUTATHIONE-RELATED ENZYMES: POSSIBLE PHYSIOLOGICAL INTERACTION, GENE CLONING AND ABIOTIC STRESS RESPONSE

Mohammad Anwar Hossain[1,2], Daud Hossain[3], Motiar Rohman[3], Jaime A. Teixeira da Silva[4] and Masayuki Fujita[1,*]

[1]Laboratory of Plant Stress Responses, Department of Applied Biological Science, Faculty of Agriculture, Kagawa University, Miki-cho, Kita-gun, Kagawa, Japan

[2]Department of Genetics & Plant Breeding, Bangladesh Agricultural University, Mymensingh, Bangladesh

[3]Senior Scientific Officer, Bangladesh Agricultural Research Institute, Joydebpur, Gazipur, Bangladesh

[4]Laboratory of Ornamental Floriculture, Department of Bioproduction Science,Faculty of Agriculture, Kagawa University, Miki-cho, Kita-gun, Kagawa, Japan

---

[*] Corresponding author: Masayuki Fujita, Professor, Laboratory of Plant Stress Responses, Faculty of Agriculture, Kagawa University, Miki, Kagawa 761-0795, Japan. Email: fujita@ag.kagawa-u.ac.jp

ABSTRACT

Onions are valued globally for their medicinal attributes. Flavonoids and organosulfur compounds are the two major classes of secondary metabolites found in onions that are believed to play a role as health-promoting, disease-preventing dietary supplements, including antioxidant activity. Onion (*Allium cepa* L.) bulbs contain high levels of glutathione *S*-transferases (GSTs), glyoxalase I (Gly I), glyoxalase II (Gly II) and alliinase activities that are associated with quality, taste and aroma of foodstuffs. GSTs are involved in the intracellular detoxification of xenobiotics and endogenous toxins whereas Gly I and Gly II detoxify methylglyoxal (MG), a toxic compound produced under normal physiological conditions. Alliinase, on the other hand, is involved in the metabolism of sulfur compounds derived from glutathione (GSH). The interaction of onion bulb GSTs, separated by DEAE-cellulose chromatography, with sulfur compounds has been investigated. Two non-physiological compounds, *S*-hexyl GSHand *S*-butyl GSH, strongly inhibited the 1-chloro-2,4-dinitrobenezene (CDNB)-conjugating activity of GSTa, GSTb and GSTe. However, physiological sulfur compounds, *S*-methyl GSH, *S*-propyl GSH, *S*-lactoyl GSH and *S*-ethyl-L-cysteine sulfoxide, have small or almost no inhibitory effects. Therefore, onion sulfur compounds might have the least possibility to be the substantial physiological counterparts of onion GSTs. On the other hand, the activities of GSTc, GSTd and *Ac*GSTF1 are strongly inhibited by flavonoids, quercetin, luteolin, apigenin and kaempferol. The ethylacetate extract of onion bulbs contains quercetin-4′-glucoside as a major inhibitory substance, whose strong inhibitory effects on GSTc, GSTd and *Ac*GSTF1, as well as its high concentration in onion bulbs, indicate that it is a physiological counterpart of dominant GSTs in onion bulbs. GSTe was slightly inhibited by quercetin, but the activity showed a significant seasonal variation between onion bulbs harvested in autumn and spring, suggesting that this GST isoform might play a role in abiotic stress tolerance through an unknown function. We purified and characterized onion GST and Gly I proteins and studied their expression pattern under various abiotic stresses. Three types (short-, medium- and long-) of onion Gly I cDNA were also isolated, cloned and sequenced. However, only short-type Gly I cDNA showed enzymatic activity. The two other types of onion Gly I had no enzymatic activity, although their presence indicates that they might have other physiological functions. The evolutionary relationship showed that medium- and long-type Gly I evolved from the gene duplication of short-type Gly I sequences. In this chapter, we will provide an overview of the physiologicalattributesof GST, Gly I, Gly II and allinase in onion bulbs and cultured cells. The possible relationship among these enzymes as well as their physiological inhibitors and in relation to stress tolerance will also be discussed.

1. INTRODUCTION

In recent years there has been a remarkable increase in scientific research reports dealing with oxidative stress and the formation of reactive oxygen species (ROS), linked to the leading cause of various chronic and lifestyle diseases in modern times. Regarding this, vegetables and fruits rich in phytochemicals that can aid in the prevention of diseases are of major interest to both the scientific community and the general public (Kaur et al., 2007; Hossain et al., 2007a). Among vegetables, onions (*Allium cepa* L.) are valued globally for their medicinal attributes. Onion is a rich source of a variety of organosulfur compounds such as cepaenes and thiosulfinates (Dorsch & Wagner, 1991; Goldman et al., 1996; Lancaster et al., 2000), the large class of flavonoids including quercetin and kaempferol (Dorant et al., 1994; Aoyama &Yamamoto, 2007; Roldán-Marín et al., 2009; Rohman et al., 2009a,2009b), and pigments such as anthocyanins (Fitzpatrick et al., 1993; Gorinstein et al., 2008; Rodrigues et al., 2009). Organosulfur compounds and flavonoids are believed to possess anti-inflammatory, antioxidant, anti-allergic, anti-microbial, and anti-thrombotic activity (Middleton & Kandaswami, 1993; Hertog et al., 1993). They also act as enzyme inhibitors, precursors of toxic substances, and pigments and light screens (McClure, 1986; Harborne et al., 1993; Block et al., 1997). Other evidence suggests that most of the compounds in onions are believed to play a role as health-promoting and disease-preventing dietary supplements (Malaveille et al., 1996; Gazzani et al., 1998).

Although several studies have demonstrated the importance of secondary metabolites, including organosulfur compounds and flavonoids, as plant defense compounds (Tailor & Grotewold, 2005) many of them exert toxic effects on other organisms (Marrs, 1996; Alfenito et al., 1998) and even to the cells that produce them. Therefore, to avoid self toxicity, plants employ different self-defense systems to eliminate or modify the toxicity of these compounds. Among the defense mechanisms, enzymatic defense is the most important one. In this context, glutathione *S*-transferases (GSTs; EC 2.5.1.18) and glyoxalase I (Gly I; EC 4.4.1.5) are of great importance since they utilize reduced glutathione (GSH). GSTs are involved in the intracellular detoxification of xenobiotics and endogenous toxins and in their vacuolar sequestration (Mannervik & Danielson, 1988; Edwards et al., 2000; Mueller et al., 2000; Dixon & Edwards, 2010). Beside detoxification, they also play an important role in other physiological functions such as GSH peroxidase activity (Bartling et al., 1993; Mannervik & Danielson,

1988) and intracellular binding and transport of phytochemicals (Edwards et al., 2000). Gly I plays a key role in the detoxification of toxic methylglyoxal (MG) in living cells through the conversion of MG to S-D-lactoylglutathione (SLG), which is then converted to D-lactic acid by glyoxalase II (Gly II; EC 3.1.2.6) (Racker, 1951). Both Gly I and Gly II play important roles under stress conditions since their levels have been reported to increase in response to various abiotic stresses (Espartero et al., 1995; Yadav et al., 2005a, 2005b; Hoque et al., 2008; Kumar & Yadav, 2009; Lin et al., 2010; Hossain & Fujita, 2009; Hossain et al., 2009, 2010, 2011b).

Garlic, onion and related *Allium* species have a characteristic smell and taste which are associated with alliin lyase (alliinase; EC 4.4.1.4). Alliinase is present in *A. cepa* bulbs at high levels, representing up to 6% (w/w) of the soluble protein in the bulb (Nock & Mazelis, 1987). The enzyme plays an important role in secondary metabolism of sulfur compounds derived from GSH (Tobkin & Mazelis, 1979; Nock & Mazelis, 1989; Wang et al.,2011a). It catalyzes the conversion of *S*-alk(en)yl-L-cysteine sulfoxides (flavour precursors) into corresponding thio-sulfinates that are responsible for flavour and taste. The roles of the flavour compounds as well as alliinase also include defense against pests and predation, storage and transportation of carbon, nitrogen and sulfur(Lancaster & Boland, 1990), and prolongation of shelf-life of agricultural crops (Jones et al., 2004).

GSTs are thought to play vital roles in the diversity of stress physiologies. A wide range of biotic and abiotic factors including herbicides, heavy metal, pathogen attack, ethylene, ozone, auxin, salicylic acid and hydrogen peroxide induce plant GST expression (Marrs, 1996). Environmental stresses like osmotic stress, low temperature, salinity and cadmium also induce GST expression in plants (Boot et al., 1993; Galle et al., 2008; Marrs & Walbot, 1997; Fujita & Hossain, 2003a, 2003b; Hossain & Fujita, 2002; Hossain et al., 2006a, 2006b). On the other hand, a variety of sulfur compounds and flavonoids have also been reported to inhibit the activities of plant GSTs (Mueller et al., 2000; Cummins et al., 2003; Hossain et al., 2007b, 2008). However, there is currently little information on endogenous substances of onion that inhibit GST activity. Importantly, the regulation of Gly I activity due to different abiotic and biotic stresses has recently been demonstrated. Gly I from tomato and *Brassica* were shown to be upregulated under salt, water and heavy metal stresses (Espartero et al., 1995; Veena et al., 1999). Treatments that stimulate cell growth, including phytohormones (auxins, cytokinins, etc.) and blue light also increased Gly I activity (Chakravarty & Sopory, 1998). Conversely, inhibition of cell growth resulted in lower levels of Gly I activity (Deswal et al., 1993; Paulus et al., 1993). Very recently, the importance of the glyoxalase pathway in stress tolerance in plants has come to light due its interaction with antioxidant defense pathways and systems through a multifunctional redox metabolite like GSH (Yadav et al., 2005a, 2005b; Hoque et al., 2008; Hossain & Fujita, 2009, 2010, 2011; Kumar & Yadav, 2009; El-Shabrawi et al., 2010; Hossain et al., 2009, 2011a, 2011b). In this chapter, we will provide an overview of GST, Gly I, Gly II and allinase metabolism in onion bulbs and cultured cells and their interaction with their physiological counterparts. Further, we will discuss the progress made over the last few years in our understating of the genetics, evolution, physiology and molecular biology of two important thiol-dependent enzymes, namely GST and Gly I, and their differential regulation and response under various abiotic stress conditions.

2. Mature Onion Bulb Expresses High GSTs, Gly I and Alliinase Activities

Crude enzyme solutions were prepared from pumpkin (fruit parenchyma tissue), radish (root), carrot (root), sweet potato (tuber), potato (tuber), broccoli (branchlet with floret), cabbage (leaves), mungbean seedlings and mature onion bulb and GST, Gly I and alliinase activities were determined by standard methods (Hossain et al., 2007a). Mature onion bulbs showed significantly higher specific activities of GST (648 nmol $min^{-1}mg^{-1}P$), Gly I (4540 nmol $min^{-1}mg^{-1}P$) and alliinase (2069 nmol $min^{-1}mg^{-1}P$) than those of other vegetable crops (Figure 1) although the activities of the three enzymes were detected in all the vegetables examined. A large number of biochemical studies have supported that many higher plants have GST (Hahn & Strittmatter, 1994; Lopez et al., 1994; Edwards & Dixon, 1991; Jain et al., 2010), Gly I (Thornalley, 1990; Deswal & Sopory, 1991, 1998; Veena et al., 1999; Jain et al., 2002; Tuomainen et al., 2011) and alliinase activity (Tobkin & Mazelis, 1979; Lancaster et al., 2000; Jones et al., 2004; Wang et al., 2011a). Recently, we have been conducting studies on GSTs and Gly I over a range of crop varieties under various abiotic stress conditions (Fujita et al., 1994, 1998; Hossain & Fujita, 2002, 2009, 2010; Fujita & Hossain, 2003a, 2003b; Hossain et al., 2006a, 2009, 2010, 2011a, 2011b; Hasanuzzaman et al., 2011a, 2011b).

The higher level of GST activity in onion bulbs suggests that they are capable of detoxifying toxins in the presence of sufficient GSH. However, high GST activity in onion bulbs might be correlated with high concentrations of organosulfur compounds in the soluble extract. On the other hand, the existence of Gly I activity in vegetables indicates their tolerance mechanisms against MG, which occurs endogenously through different physiological processes and is toxic to cells as it arrests growth (Szent-Gyorgi et al., 1967), reacts with proteins and nucleic acids (Thornalley, 1996), and inactivates the antioxidant defense system (Martins, 2001; Hoque et al., 2010; Hossain et al., 2011b). The high alliinase activity in onion indicates that the enzyme is needed to convert *S*-alk(en)yl-L-cysteine sulfoxides into corresponding thio-sulfinates that are responsible for onion's characteristic flavor and taste. The higher allinase activity might improve shelf-life of onion since mild-flavored onions were reported to have poorstorage properties (Jones et al., 2004).

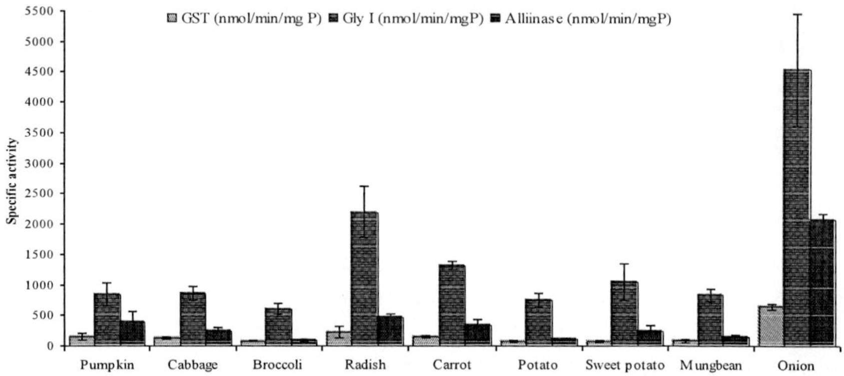

Figure 1.The specific activity of GST, Gly I and alliinase in soluble extracts of different vegetable crops. Results were obtained from three independent experiments and bars indicate the standard error (SE).

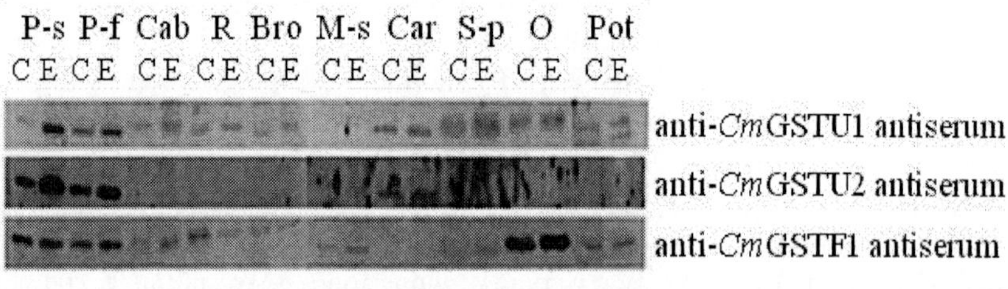

Figure 2.Detection of putative GSTs of various vegetable crops by Western blot analysis. Anti-pumpkin GST antisera were used as primary antisera. Each of the vegetables (P-s: pumpkin seedling; P-f: pumpkin fruit; Cab: cabbage; R: radish; Bro: broccoli; M-s: mungbean seedling; Car: carrot; S-p: sweet potato; Pot: Potato and O: onion) was incubated for 24 h with (E) and without (C) ethanol vapour (adapted from Hossain et al., 2007a).

3. Cross Reactivity of Putative Onion GSTs with Anti-pumpkin GSTs Antisera

The highest specific GST activity level in onion extract is unexpected since the quantity of total GSH in onions has been reported to be less than that in other vegetables (Nakagawa et al., 1986). To investigate the presence of GST in a different manner, we directly compared the putative GST band of onion with that of other vegetables by Western blot analysis. Here, the antisera of pumpkin GSTs (respective anti-rabbit sera) were used as primary antisera (i.e., anti-*Cm*GSTU1, anti-*Cm*GSTU2 and anti-*Cm*GSTF1 antisera). An important characteristic of plant GSTs is that most of them are induced by ethanol (Cummins et al., 1997; Dixon et al., 1998; McGonigle et al., 2000; Fujita & Hossain, 2003b). Therefore, we also treated the plant materials with ethanol vapour and compared its effect with the control. The results (Figure 2) show that each anti-pumpkin GST antiserum recognized some protein bands in most of the vegetable extracts that could be referred to as putative GSTs as they were located in a position similar to those of pumpkin GST bands in SDS-PAGE. The results also revealed that ethanol vapor considerably induced the expression of GSTs in pumpkin seedlings and fruit, cabbage, sweet potato and onion. When treated with anti-*Cm*GSTU2 antiserum, the putative GST bands were only detected in pumpkin seedlings and fruit. No clear GST bands for either controls or ethanol treatment for any of the other vegetables were detected. The anti-*Cm*GSTF1 antiserum could detect GST bands in most of the vegetable extracts and the antiserum cross-reacted strongly with the putative GST of onion as its band was found to be much thicker than that of other vegetables.

4. Purification and Partial Characterization of Onion GSTs

We partially purified onion bulb GSTs by DEAE-cellulose column chromatography and the GST activities of all DEAE fractions were estimated according to the model substrate CDNB (Rohman et al., 2009a, 2009b). Five highly active GST peaks were eluted at approximately 43, 65, 106, 117 and 157 mM KCl (Figure 3), designated as GSTa, GSTb, GSTc, GSTd and GSTe, respectively. Among them, GSTa and GSTb, accounting for only 2.0 and 1.1% of the total activity, were termed minor GSTs, and GSTc, GSTd and GSTe, accounting for 27.7, 35.2 and 33.6% of the total activity, were termed dominant GSTs. Next, we directly applied the peak fraction of the five GSTs separately onto an affinity column of *S*-hexylglutathione-agarose. Fractions of each GST were collected and applied to SDS-PAGE electrophoresis for silver staining. The comparative quantitative determination by silverstaining showed that the amount GSTa and GSTb were not as low as their activity (Figure 4). Since the model substrate CDNB used in this experiment is not a physiological substrate, GSTa and GSTb might fully express to another physiological substrate.In general, GSTs are 25-30-kDa proteins. From our results, we found that GSTc, GSTd and GSTe (dominant GSTs) are 27-kDa proteins, and GSTa and GSTb have a higher molecular weight than the dominant GSTs, suggesting that GSTa and GSTb might be different from dominant GSTs (Rohman et al., 2009b).

In order to investigate the cross reactivity of onion bulb GSTs with anti-*Cm*GSTF1 (pumpkin phi type GST) antiserum, we conducted Western blot analysis of the active fractions. The anti-*Cm*GSTF1 antiserum detected an intensive band of GST signals for GSTc and GSTd, suggesting that they might consist of one or two subunits reactive to the antiserum but GSTa, GSTb and GSTe were non-reactive to the anti-serum(Rohman et al., 2009a). As GSTe is clearly separated from the other dominant GSTs (GSTc and GSTd) and has higher activity, we tried to determine the genetic structure of the responsible gene for that protein. We purified GSTe by DEAE-cellulose, hydroxyapatite and *S*-hexyl glutathione-agarose column chromatography. The final product was purified 93-fold with a yield of 3.5%. The specific activity of purified GSTe was 72.3 μmol $min^{-1}mg^{-1}$protein towards CDNB. The purified enzyme showed a single band with a molecular mass of 27-kDa on SDS-PAGE (Rohman et al., 2010a).

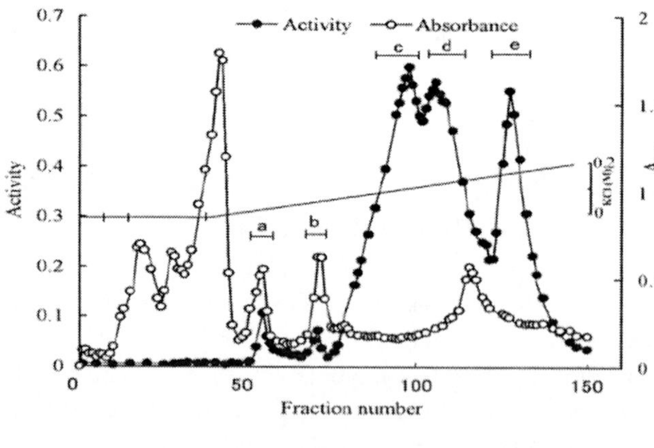

Figure 3.Typical column chromatography of DEAE cellulose of soluble proteins prepared from 150g onion bulb tissues. For each fraction, absorbance at 280 nm (○) and GST activity toward CDNB (●) were determined. Activity is expressed as μmol $min^{-1}ml^{-1}$. The fractions under the bar were pooled (adapted from Rohman et al., 2009b).

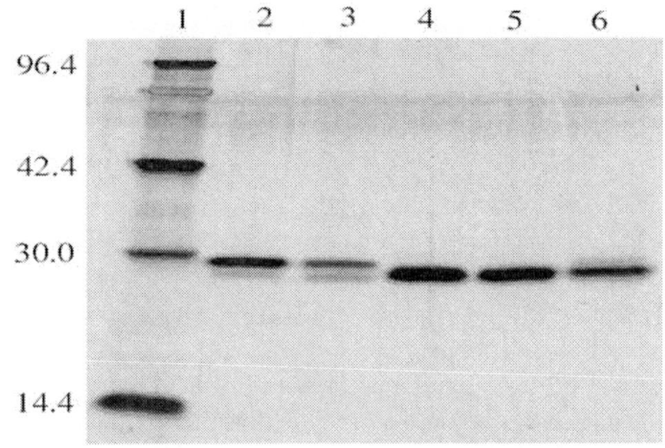

Figure 4. Silver staining of GSTs purified by an affinity column to which each peak fraction of five partially purified GSTs was applied. *Lane1*, molecular weight marker; *Lane 2*, GSTa; *Lane 3*, GSTb; *Lane 4*, GSTc; *Lane 5*, GSTd and *Lane 6*; GSTe. Each lane contains 5.1 μg of protein (adapted from Rohman et al., 2009b).

5. Molecular Cloning and Characterization of Onion GST Gene

To ascertain the genetic structure of the GST gene corresponding to the GSTc protein eluted in the DEAE-cellulose column, we immunoscreened an onion bulbλZap II cDNA library with anti-*Cm*GSTF1 antiserum. We isolated a positive cDNA clone and sequenced it. The GST cDNA consist of a820-bp nucleotide encoding a polypeptide of 212 amino acids having a predicted molecular weight of 23,724 and a predicted isoelectric point (pI) of 5.74. It has 46% identity and 61% similarity (in deduced amino acids) to *Cm*GSTF1, and we designated it as a phi-type like GST, *Ac*GSTF1 (Accession number: AB300334). The fusion protein expressed in the bacterial cells was bound with anti-*Cm*GSTF1 antiserum in Western blotting (Figure 5). Though the values of identity and similarity seem to be low to represent the high reactivity of anti-*Cm*GSTF1 antiserum to *Ac*GSTF1, the antiserum might strongly recognize some specific epitopes of *Ac*GSTF1. Previously, in our lab, a homodimer of *Cm*GSTU2 (Pugb) (pI 5.26) and a heterodimer of *Cm*GSTU2 and *Cm*GSTU3 (Pugc) (pI 6.01) were eluted at approximately 105 mM KCl in DEAE-cellulose column chromatography under the same condition (Fujita et al., 1994) suggesting that *Ac*GSTF1 (pI 5.74) is a subunit of GSTc (106 mM KCl). Besides, to determine the inner sequence of GSTc, we directly purified GSTc with *S*-hexyl glutathione-agarose from the corresponding fractions using DEAE-cellulose chromatography (Figure 3) and then digested it with BrCN. One of the peptides obtained in the digestion had a sequence [(M)KSGEHKKPEF] identical to the part (from 35 to 45 in amino acids) of the deduced sequence from *Ac*GSTF1 cDNA. This finding strongly suggests that GSTc is a homodimer of *Ac*GSTF1 (Rohman et al., 2009a).

Additionally, we produced rabbit antiserum using purified protein (GSTe) and used asanit-GST antiserum. We immunoscreened an onion cDNA library, isolated a cDNA clone and sequence it. Importantly, we obtained a new GST gene *Ac*GSTU1 (Accession number: AB 627988), which consists of a 891-bp nucleotide encoding a polypeptide of 221 amino acids having a predicted molecular weight of 26,000 and a predicted pI of 6.24 but we could not find the target gene sequence. Based on the pI value, this nucleotide corresponds to GSTa or GSTb (Figure 3) protein eluted in the DEAE-cellulose column (Hossain & Fujita unpublished data).

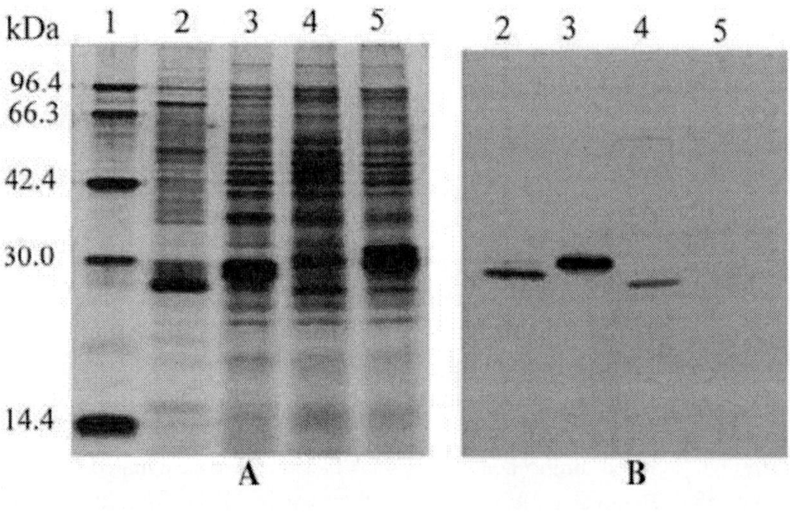

Figure 5.Silver staining (a) and Western blotting (b) of bacterial expression products for cloned GST cDNAs (*AcGSTF1, CmGSTF1* and *CmGSTU3*) and the GSTc fraction in DEAE-cellulose chromatography.*Lane 1* molecular weight marker;*Lane 2*, GSTc; *Ac*GSTF1, *Lane 4, Cm*GSTF1;*Lane 5, Cm*GSTU3. The bacterial products were expressed as fusion proteins of GSTs and anα-complementation particle ofβ-galactosidase. Each lane contains 6.5 μg protein, and the specific activity of GSTc, *Ac*GSTF1, *Cm*GSTF1 and *Cm*GSTU3 was 1,022, 4,706, 469 and 5,384 nmolmin^{-1} mg^{-1} protein, respectively (adapted from Rohman et al., 2009a).

6. Evolutionary Relationship of Onion GST Gene

To visualize the evolutionary relationship of *Ac*GSTU1 and *Ac*GSTF1 with other known GSTs, a phylogenetic tree was constructed. Based on the evolutionary relationship of deduced amino acid sequence, it was concluded that *Ac*GSTU1 is a tau type and *Ac*GSTF1 belonged to the phi type GST (Figure 6). The phylogenetic tree showed that *Ac*GSTF1 is closely related to *Zm*GST1, whereas *Ac*GSTU1 is closely related to *Cm*GSTUI. Both tau and phi type GSTs are the most common, are plant-specific and are responsive to various abiotic stresses (Zhou & Goldsbrough, 1993; Ezaki et al., 1995; Polidoros & Scandalios, 1999; Hossain & Fujita, 2002, 2010; Fujita & Hossain, 2003a, 2003b; Rohman et al., 2010a; Jha et al., 2010; Diao et al., 2011).

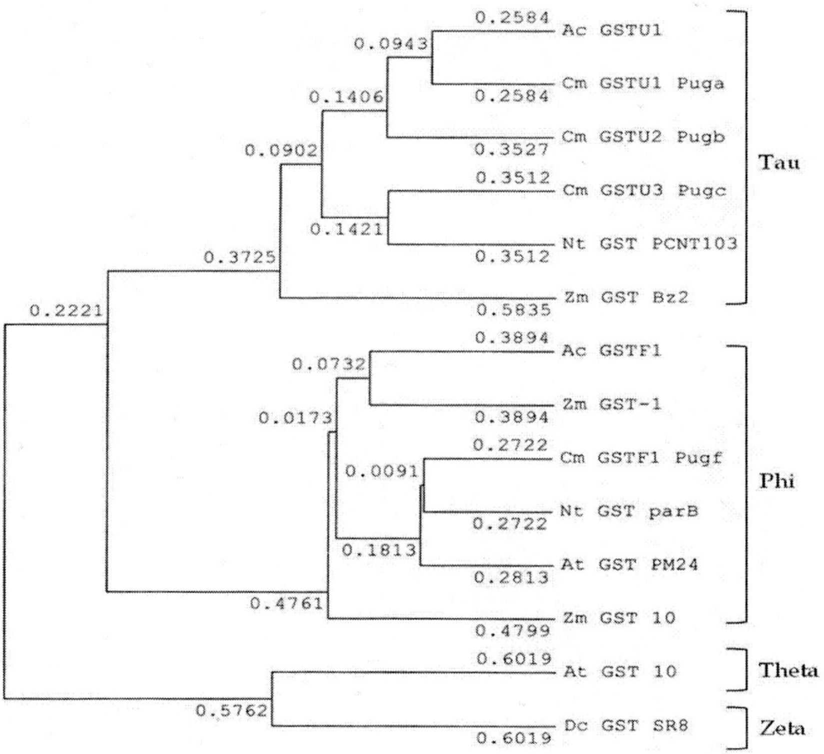

Figure 6.Phylogenetic tree showing the evolutionary relationship of *Ac*GSTU1 and *Ac*GSTF1 with other known plant GSTs. The tree was constructed using CLUSTALW software.

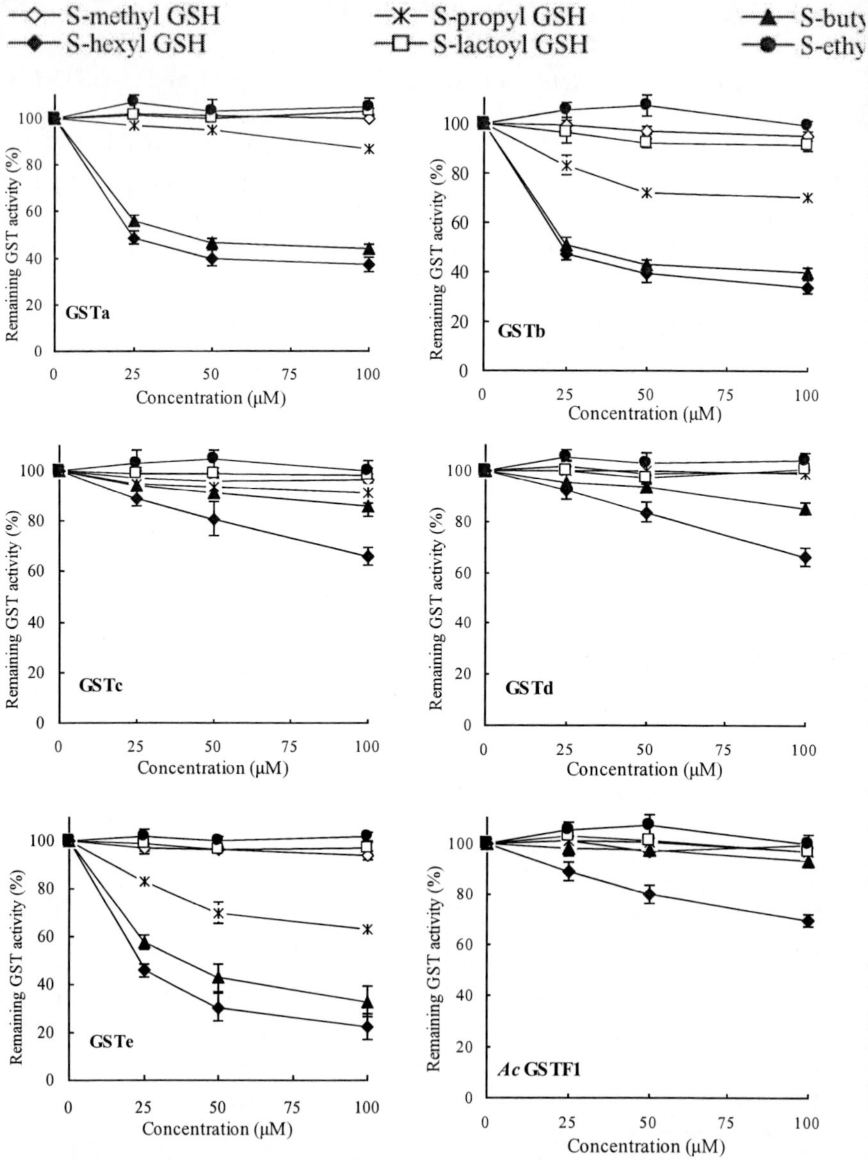

Figure 7.Inhibition of the CDNB-conjugating activity of GSTa, GSTb, GSTc, GSTd, GSTe and *Ac*GSTF1 by GSH and cysteine derivatives. Results were obtained from three independent experiments and bars indicate SE (adapted from Rohman et al., 2009a).

7. Interaction of Onion GSTs with Sulfur Compounds

The onion plant is a rich source of a variety of sulfur compounds and among them GSH and cysteine derivatives are important biosynthetics (Lancaster & Kelly, 1983; Jones et al., 2004). Some of the derivatives have been reported to inhibit pumpkin GSTs (Hossain et al., 2007b). We therefore, investigated the inhibitory potency of GSH and cysteine derivatives on the activities of GSTa, GSTb, GSTc, GSTd, GSTe and *Ac*GSTF1 towards model substrate CDNB (Figure 7). Among the compounds, *S*-hexyl GSH and *S*-butyl GSH showed strong inhibitory effects on the activities of GSTa, GSTb and GSTe. However, *S*-methyl GSH, *S*-propyl GSH, *S*-lactoyl GSH and *S*-ethyl-L-cysteine sulfoxide had small or almost no inhibitory effects. In an earlier experiment, we also found that *S*-hexyl GSH was a potent inhibitor of onion GSTs prepared from mature onion bulbs by $(NH_4)_2SO_4$precipitation followed by overnight dialysis against Tris-HCl buffer (Hossain et al., 2007a). Although *S*-hexyl GSH and *S*-butyl GSH have not yet been reported to be present naturally in plants, the existenceof *S*-methyl GSH, *S*-propyl GSH *S*-lactoyl GSH (a product of Gly I) and *S*-ethyl L-cysteine sulfoxide (a common flavour precursor of *Allium* spp.) in onion is quite natural and plausible.However, all of these physiological substances were not found to have interacted strongly with onion GSTs. These results, therefore, reveal that onion sulfur compounds have the least possibility of being potent physiological inhibitors of onion GSTs.

Table 1. Inhibition of onion GSTs by flavonoids (IC_{50} values in μM)

Flavonoids		IC_{50}			
		GSTc	GSTd	GSTe	*Ac*GSTF1
Chalcone	Chalcone	NI	NI	NI	NI
Flavanonol	Taxifolin	NI	NI	NI	NI
Flavanone	Naringenin	NR	NR	NI	NR
Flavone	Luteolin	27.0 ± 3.5	34.0 ± 2.8	NI	43.0 ± 4.5
	Apigenin	33.0 ± 2.8	32.8 ± 2.1	NI	>50
Flavonol	Kaempferol	15.5 ± 3.5	18.8 ± 3.9	NI	23.0 ± 4.3
	Quercetin	21.0 ± 3.0	20.6 ± 4.3	NI	25.5 ± 3.9

IC_{50} values were graphically determined by plotting GST activities towards CDNB as a function of inhibitor concentration. Each value for IC_{50} is the mean of three independent experiments ± SE.

NR not reached at 50% level, *NI* negligible inhibition (<10%).

8. Interaction of Onion GSTs with Flavonoids

The onion bulb is an important source of flavonoids, particularly quercetin (Arabbi et al., 2004; Moskaug et al., 2004), and a few studies have suggested that plant GSTs have a potential role as flavonoid-binding proteins (Edwards et al., 2000; Mueller et al., 2000). Flavonoids have also been reported to inhibit the activity of GSTs in petunia (*Petunia hybrida*) (Mueller et al., 2000) and in wheat (*Triticum aestivum* L.) (Cummins et al., 2003). We examined the effects of quercetin and its precursors on the activity of the dominant onion bulb GSTs, including *Ac*GSTF1 (Rohman et al., 2009a). The CDNB-conjugating activity of GSTc, GSTd and, to a lesser extent, *Ac*GSTF1 were strongly inhibited by kaempferol and quercetin (flavonols) and by luteolin and apigenin (flavones) (Table 1). Naringenin (a flavanone) exhibited moderate inhibitory potency, whereas taxifolin (a flavanonol) and chalcone showed a negligible interaction. GSTe, which was strongly inhibited by *S*-hexyl GSH, was much less sensitive to such inhibition. Mueller et al. (2000) found that a petunia GST, AN9, is highly sensitive to inhibition by flavones and flavonols compared with its sensitivity to flavanones. Similarly, Cummins et al. (2003) reported that certain phi-type GSTs from wheat are more sensitive to inhibition by flavanonols than to inhibition by flavones and flavanones. Though the binding capacity of different proteins varied for specific flavonoids, these findings suggest that there is an important metabolic correlation between GSTs and natural flavonoids. The common biosynthetic pathway of flavonoids is flavanones-flavone-flavonol or flavanones-flavanonol-flavonol, and quercetin is the most abundant flavonol in onion (Lachman et al., 2003) with strong inhibitory activity towards GSTc and GSTd as well as *Ac*GSTF1. Therefore, onion GSTs might have an important role in quercetin metabolism. Since flavonoids are cytotoxic (Ahmed et al., 1994), GSTs might play a role in stabilizing (preventing oxidation) them in the cytosol and hence aid vacuolar sequestration (Edwards et al., 2000). However, the possibility that GSTs facilitate the delivery of flavonoids to specific compartments is still under study. The quercetin precursors, kaempferol, luteolin and apigenin, strongly inhibited the activity of the GSTs, but these flavonoids are rarely found in onion. Moreover, quercetin is found both in free and glucosidic forms in onion (Rhodes & Price, 1996).

9. Purification and Detection of Physiological Inhibitors of Onion GSTs

We extracted low molecular inhibitory substances from mature onion bulbs with 50% methanol, *n*-hexane, ethyl acetate (EtOAc) and finally with water following a standard method (Rohman et al., 2009a, 2009b). The presence of inhibitory substances in *n*-hexane, EtOAc and water fractions was tested by inhibition of dominant GSTs and *Ac*GSTF1. Water and EtOAc fractions showed strong inhibitory effects on the GSTs except for GSTe, indicating that both of the fractions contain potent inhibitory substances for onion GSTs. Next, we fractionated the water and EtOAc fractions by a Seppak Vac C18 column with 20, 45, 65 and 100% methanol and the major inhibitory substance in GSTc, GSTd and *Ac*GSTF1 fractions was eluted with 45% methanol. The inhibitory substances of the fraction were further purified by thin layer chromatography (TLC) and high performance liquid chromatography (HPLC) and their structural determination was performed by nuclear magnetic resonance (NMR) spectroscopy and fast atom bombardment mass spectrometric (FAB-MS) spectroscopy. The structures of the inhibitory substances were confirmed as quercetin-4′-glucoside and quercetin-3, 4′-diglucoside in the EtOAc and water fractions, respectively (Figure 8), based on comparison with the data of Fossen et al. (1998).

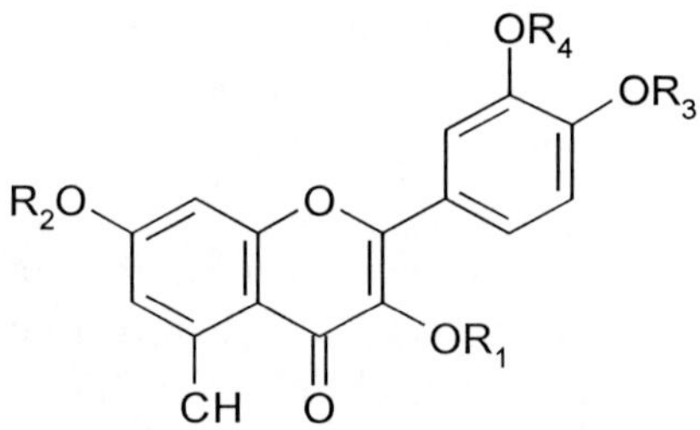

R_1,R_2,R_4=H; R_3=glu.: Quercetin-4′-glucoside

R_2,R_4=H; R_1,R_3=glu.: Quercetin-3,4′-diglucoside

Figure 8.Structures of physiological inhibitors of onion GSTs.

10. Interaction of Onion GSTs with their Physiological Inhibitors

The inhibitory expression of quercetin-4′-glucoside and quercetin-3,4′-diglucoside on the activity of different onion bulb GSTs was investigated (Rohman et al., 2010b). Both of the compounds showed the strongest inhibitory effect on GSTd followed by GSTc (Figures9A& B). The activity of GSTe was inhibited to a lesser extent that the activity of other GSTs. These results suggest that onion GSTc, GSTd and *Ac*GSTF1 are involved in the metabolismof quercetin-4STglucoside and quercetin-3, 4ucdiglucoside as well as its aglycone (quercetin).

High inhibitory potency and *in vivo* concentration of a certain compound are important factors for it to be a physiological inhibitor of an enzyme. The strong inhibitory effects of quercetin-4′-glucoside on GSTc and GSTd as well as on *Ac*GSTF1 (IC_{50}s: 9.5, 7.5 and 11.2 μM, respectively) along with its high concentration (226 μM) in onion bulbs indicate that quercetin-4′-glucoside is a physiological inhibitor of the dominant onion GSTs. According to the proposal of Walbot et al. (2000), plant GSTs act as non-enzymatic carrier proteins for plant natural compounds before transportation into the vacuole. From our study, the IC_{50} of quercetin-3,4′-diglucoside on GSTc and GSTd was 76.3 and 69.3 μM, respectively and its concentration in onion bulbs was 119 μM suggesting that the high levels of onion GSTs might function as carriers/ligands to deliver most of the cellular quercetin-3,4′-diglucoside into the vacuole.

Onion bulbs exhibited higher GST activity than other vegetable crops (Hossain et al., 2007a) and several research groups have detected GST activity in different organs of onion plants (Lamoureux & Rusness, 1980; Schröder & Stampfl, 1999; Fatima & Ahmed, 2005; Tabrez & Ahmad, 2011). We investigated the activity and accumulation of GST in different parts of healthy onion plants and in onion calli induced from onion seeds cultured on Murashige and Skoog solid medium containing 1 μg ml^{-1} 2, 4-D and 0.1 μg ml^{-1} kinetin at 25°C(Hossain & Fujita, 2009).

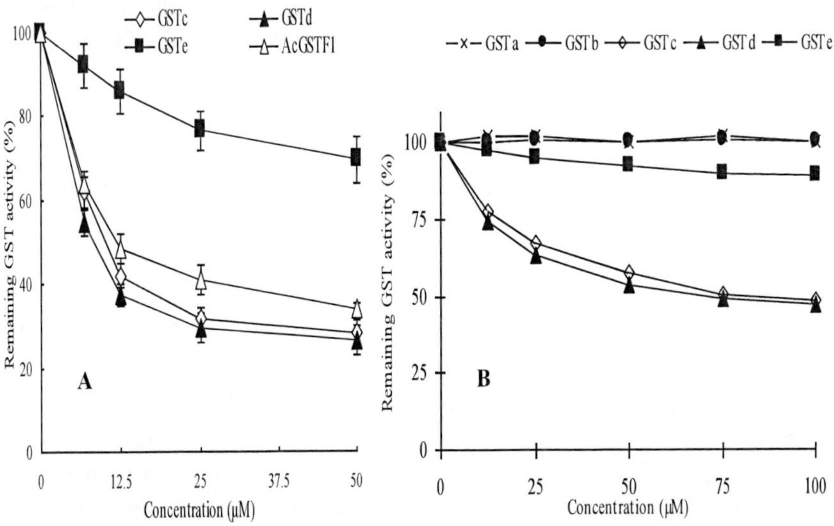

Figure 9. Inhibitory expression of quercetin-4′-glucoside (A) and quercetin-3,4′-diglucoside (B) on CDNB conjugating activities of onion GSTs. Results were obtained from three independent experiments and bars indicate SE (adapted from Rohman et al.,2009a, 2009b).

11. Expression of GST in Different Organs in Healthy Onion Plants and Callus

Root tissues had the highest GST activity followed by mature bulbs, calli, and young bulbs (Figure 10). In contrast, young and mature leaves had lower GST activity. The GST activity was higher in mature organs than in young ones. The accumulation of GST isoforms (GSTe and *Ac*GSTF1) – checked by Western blotting –varied greatly among different plant parts and calli. The accumulation of GSTe was highest in mature onion bulbs followed by calli, young bulbs, roots and mature leavesbut was not detected in young leaves. *Ac*GSTF1 was detected in all organs. Roots accumulated the highest level of *Ac*GSTF1 followed by mature bulbs, callus, young bulbs and mature leaves. *Ac*GSTF1 in young leaves was barely present (Rohman et al., 2010a). GST accumulation may vary with gene member in tissues or spatially and temporally during plant growth and development. In *Arabidopsis*, one GST homolog, *At103-1a*, expressed constitutively in the green part of the plant but the transcript was barely detectable in roots (van der Kop et al., 1996). However, expression of another GST gene, *At103-1b*, could be detected in both green plant parts and in roots. Several GST isoforms (*At*GSTF2, *At*GSTF8 and*At*GSTU19) were specifically expressed during the formation of the early root epidermis of *Arabidopsis* (Mang et al., 2004).

12. Seasonal Variation in the Activities of Dominant Onion Bulb GSTs

Onion bulb GST isoforms (GSTa, GSTb, GSTc, GSTd and GSTe) isolated from bulbs harvested in autumn and springwere separated by DEAE-cellulose chromatography. In onion bulbsfrom both seasons, the five peaks of GST activity (GSTa, GSTb, GSTc, GSTd and GSTe) were eluted at approximately 43, 65, 106, 117 and 157 mM KCl (Figure 11). The individual activity of GSTa, GSTb, GSTc and GSTd was similar in onion bulbsfrom both seasons. However, the activity of GSTe in autumn was higher than that in spring. The seasonal variation in GSTe activity might have a biological significance. We assume that higher activity of GSTe in autumn might be involved in abiotic stress toleranceduring cultivation.

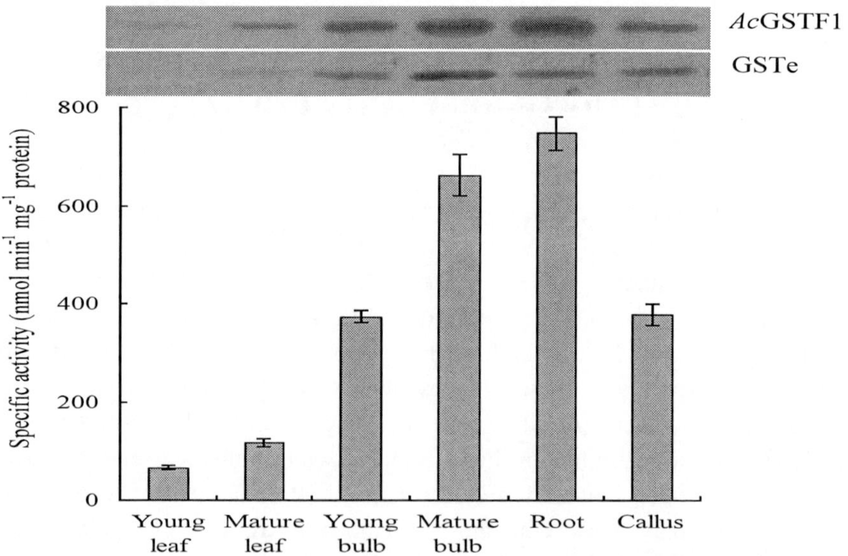

Figure 10.Expression of GST activities in different parts of onion plant and callus. Bar graph shows the GST activity towards CDNB in soluble protein extracts from young leaves, mature leaves, young bulb, mature bulb, root and callus, and Western blotting results show the accumulation of GSTeand *Ac*GSTF1 in different organs of onion

plants and callus. Each lane contained 30 μg of the protein. Results were obtained from two independent experiments and bars indicate SE (adapted from Rohman et al., 2010a).

13. Abiotic Stress Induces Differential Regulation of Onion GST

Abiotic stresses invariably induce oxidative stress and lead to oxidative damage (Fujita & Hossain, 2003a, 2003b; Hossain & Fujita, 2010, 2011; Hossain et al., 2006a, 2010, 2011a, 2011b; Hanuzzaman et al., 2011a, 2011b). Plants employ several defense mechanisms to counteract stress-induced oxidative damage (Mittler et al., 2004; Hossain & Fujita, 2011; Hossain et al., 2011b). GST is known to function as a pivotal component in plant defense systems and catalyzes the reduction reaction of toxic organic hydroperoxides by using GSH as a co-substrate or coenzyme (Dixon & Edwards, 2010). In plants, many GSTs are induced by various abiotic stresses such as salinity, dehydration, low and high temperatures, heavy metals, pesticides, ozone, pathogens and plant hormones (Hossain et al., 2006b and reference therein). Therefore, plant GSTs are thought to play vital roles in a diversity of stress physiology through some other functions as well as GST-mediated detoxification. Among the oxidative defense enzymes, the importance of GSTs in intracellular redox state has been reported (Roxas et al., 1997; Polidoros & Scandalios, 1999; Dixon et al., 2002; Dixon & Edwards, 2010; Foyer & Noctor, 2011). We investigated the physiological response of two GSTs (GSTe and *Ac*GSTF1) and their induction in onion calli subjected to salinity (200 mM NaCl), osmotic (10% PEG), low temperature (4°C) and heavy metal (1 mM $CdCl_2$) stress (Rohman et al., 2010a). All of the stresses caused a significant increase in GST activity in soluble extracts within a short period (72 h). Osmotic stress sharply increased (2.2-fold) GST activity compared to non-treated callus, followed by salinity (1.8-fold), heavy metal (1.6-fold) and low temperature (1.5-fold) stresses (Figure 12). A few recent studies also showed a significant increase in GST activity in response to salt, drought, heavy metal, and H_2O_2 (Hu et al., 2009; Haluškováet al., 2009; Hossain & Fujita, 2010, Hossain et al., 2010, 2011a; Wang et al., 2011b; Hasanuzzamn et al., 2011a, 2011b). We also checked the induction pattern of GSTe and *Ac*GSTF1 through Western blotting. The accumulation of GSTeand *Ac*GSTF1 varied greatly in response to different abiotic stresses. Salt and heavy metal stresses accumulated more GSTe protein while osmotic stress showed less accumulation.*Ac*GSTF1 protein accumulated more in response to all the abiotic stresses while osmotic stress showed the highest induction indicating its contribution to increased GST activity in onion callus. Increased accumulation of both GSTs under saline and heavy metal stresses indicates that both GSTs are responsible for increasing GST activity under these stresses. Based on our findings, we therefore, suggested that GSTeand *Ac*GSTF1 are differently responsive to different stresses in onion callus. Three pumpkin GSTs, *Cm*GSTU1, *Cm*GSTU2 and *Cm*GSTF1 were also reported to be differently expressed in response to different stresses (Hossain & Fujita, 2002; Fujita & Hossain, 2003a, 2003b; Hossain et al., 2006a). Very recently, Jain et al. (2010) showed that in response to rice seedlings treated with different abiotic stresses (desiccation, salt and cold), at least 20 GST genes were differentially expressed under at least one of the abiotic stress conditions. Induction of GST protein was also observed in an osmotic stress-tolerantrice (*Oryza sativa* cv. Zhonghua 8) cultivar (Zhang & Komatsu, 2007) and in a halophyte C_4 plant (*Aeluropus lagopoides*) salt stress conditions (Sobhanian et al., 2010). Higher GST protein expression in response to salt stress was also observed in a salt-tolerant barley(*Hordeum vulgare*cv. Morex) genotype (Witzel et al., 2009). Thus, the consistent results found in our study demonstrate that the expression level of GST is enhanced under various stress conditions, implying that *de novo* GST synthesis is required for plant tolerance to those stresses and that GST plays a pivotal role in plant defense responses. Considering the presence of alcohols/aldehydes and lipid peroxidation products, including α,β-unsaturated aldehydes, in the plant body, plants GSTs might play an important physiological role in the detoxification of lipid peroxidation products and similar reactive compounds of endogenous and exogenous origin. Therefore, though the inducible responses of GSTe and *Ac*GSTF1 to stress were different, this study might have implicated that they have important roles in stress tolerance.

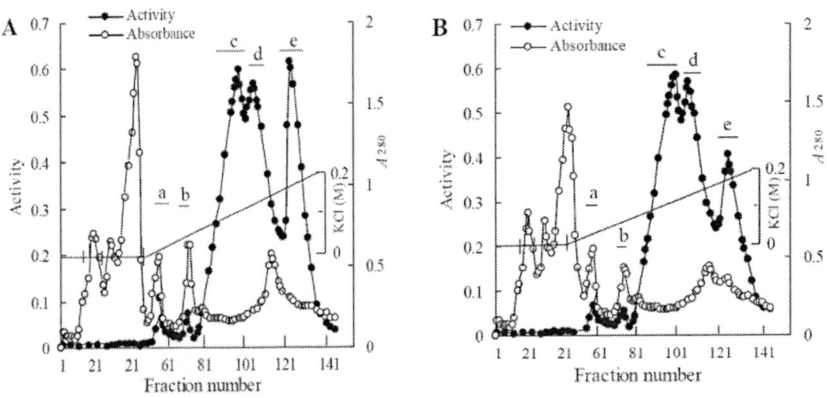

Figure 11.Seasonal variation in the GST activity of onion bulbs harvested in autumn (A) and spring (B). The GSTs were separated by DEAE-cellulose chromatography. For each fraction, absorbance at 280 nm (○) and activity towards CDNB (●) were determined. The activity was expressed as μ molmin^{-1}ml^{-1}. Bars indicate the highly active fractions of each GST isoform. The letters show the convenient names of the GSTs: a, GSTa; b, GSTb; c, GSTc; d, GSTd and e, GSTe.

14. Purification and Molecular Characterization of Onion Gly I

A wealth of recent studies in plants have demonstrated that enhancement of MG level and induction of Gly I is a common consequence of various abiotic and biotic stresses in plants and that the glyoxalase pathway is the main detoxification pathway of MG in plants (Espartero et al., 1995; Chen et al., 2004; Yadav et al., 2005a, 2005b; Hossain et al., 2009; Kumar & Yadav, 2009; Hoque et al., 2008, 2010). Therefore, to explore further insights into stress-induced regulation of the glyoxalase pathway and to identify and characterize Gly I protein and gene, we selected onionbulbs which had the highest Gly I activity compared to various plant species and plant organs (Hossain et al., 2007a). We tried to purify Gly I from onion bulbs to clarify its molecular properties. Gly I was highly purified by DEAE-cellulose, hydroxyapatite, and *S*-hexylglutathione-agarose column chromatography. The specific activity of the purified Gly I protein is the highest (356 μmol min^{-1} mg^{-1} protein) reported to date in plants and the final product was purified 184-fold in purity, with a yield of 4.18%. The purified enzyme gave a single major band on the gel stained by silver and its relative molecular mass was estimated to be approximately 25,000 (Hossain & Fujita, 2009). This apparent relative molecular mass of onion Gly I is slightly lower than that reported for Gly I purified from other plants (27,000 for *Brassica* seedlings, and two isozymes of 20,000- and 29,000 for soybean cell suspension) and closer to animal Gly I, which ranges from 21,000 to 24,000 (Thornalley, 1990; Deswal & Sopory, 1991; Paulus et al., 1993). It is important to mention that previously reported Gly I purification from other plant species (Deswal & Sopory, 1991; Paulus et al., 1993) was based on two-step purification procedures and in most of cases *S*-hexylglutathione-agarose affinity chromatography was used directly before the separation of GST and Gly I protein. *S*-hexylglutathione-agarose resin has the ability to bind both Gly I and GST proteins. Therefore, GST protein remains as a contaminant with purified Gly I, and the purified Gly I protein showed two bands on SDS-PAGE analysis (Deswal & Sopory, 1991; Paulus et al., 1993). Theantibody produced by purified Gly I identified both Gly I and GST cDNA (Skipsey et al., 2000),and also suggested that purified Gly I must have been contaminated by GST protein, but our four-step purification was effective at increasing purity by avoiding GST contamination (Hossain & Fujita, 2009).

15. Production of Antibody against Gly I and its Quantitative Analysis

We produced rabbit antiserum using purified Gly I. Western blot analysis showed a high specificity of this antibody for Gly I protein and no cross-reactivity between anti-Gly I antibody and onion GST (*Ac*GSTF1), suggesting that our purified Gly I was not contaminated by GST (Hossain & Fujita, 2009). The organ specificity of Gly I in onion plants was examined by Western blotting (Figure 13). Gly I protein was present at high concentrations in the roots, young bulbs, mature bulbs, and mature leaves, the highest concentration being in the mature bulbs.Earlier work in plants had suggested that Gly I levels correlated with the cell division status of the cells, but we found that Gly I was present in the mature bulbs, mature leaves, and flowers (non-dividing parts), indicating that it has some other role too. Recently, its role under abiotic and oxidative stresses has come to light because of the interaction of MG and ROS detoxification pathways in plants (Veena et al., 1999; Espartero et al., 1995; Yadav et al., 2005a, 2005b; Singla-Pareek et al., 2006, 2008; Hossain et al., 2009, 2010, 2011a, 2011b; Hossain & Fujita, 2009, 2010, 2011; Hoque et al., 2008, 2010; Hasanuzzaman et al., 2011a, 2011b).

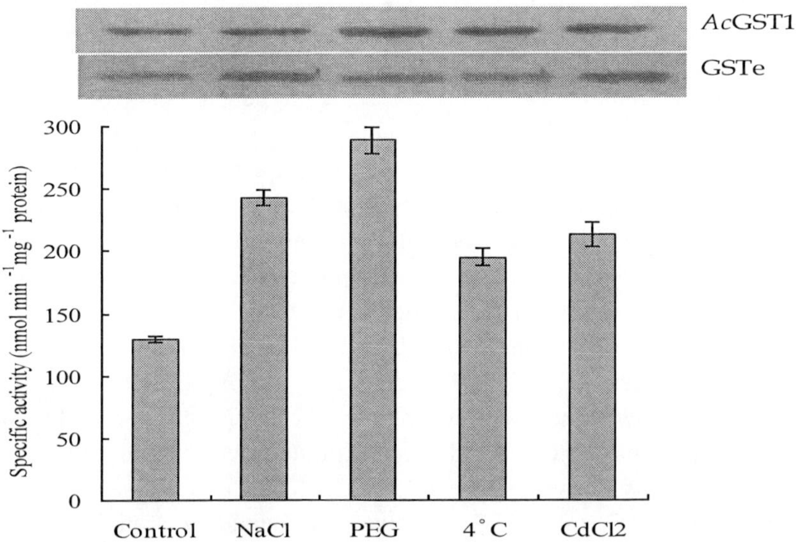

Figure 12.Effects of various stresses on the expression of GST activity in onion calli. Bar graph shows GST activity towards CDNB in soluble protein extracts under stresses: salinity (NaCl, 200mM), osmotic (PEG, 10%), low temperature (4°C) and heavy metal ($CdCl_2$, 1 mM). Western blotting results show the induction of GSTeand *Ac*GSTF1 under the stresses. Each lane contained 20 μg of soluble protein. The experiment was repeated three times and typical Western blot results are from one experiment (adapted from Rohman et al., 2010a)

16. Molecular Cloning and Characterization of Onion Gly I Gene

We constructed a λ ZAP II-cDNA library using mRNA prepared from onion bulbs and performed immunoscreening with anti-Gly I antiserum. We isolated an onion Gly I cDNA, cloned and sequence it. The onion Gly I cDNA (Accession number: AB630177) consisted of 877-bp nucleotides encoding a polypeptide of 187 amino acids having a predicted molecular weight of 21,000 and a predicted pI of 5.15. Based on amino acid number it is categorized as short-type Gly I. To determine whether this cDNA encodes Gly I protein, we expressed the cDNA encoded as a fusion protein of β-galactosidase in XL1-blue cells in the presence of 1 mM IPTG. The overexpressed *Gly I* showed 5,000-fold higher Gly I activity than the non-transformed cells, indicating that the inserted cDNA encodes Gly I. On Western blotting, the fusion protein expressed in the bacterial cells was bound with anti Gly I antiserum, indicating again that the cDNA was that of Gly I (Hossain & Fujita, 2009). Further, we immunoscreened an onion λ ZAP II-cDNA cDNA library by using *Cm*GSTF1 antiserum. Through three rounds of screening eight positive clones were obtained and sequenced. From sequencing data three short (S), two medium (L) and three long (LL) putative Gly I sequences were obtained. The molecular characteristics of onion Glyox S cDNA were the same as described previously. Onion Glyox L cDNA (*Allium cepa* Glyox homolog 1; Accession number: AB627989) consists of a 1070-bp nucleotide encoding a polypeptide of 294 amino acids having a predicted molecular weight of 33,174 and a predicted pI of 5.88. Onion Glyox LL cDNA (*Allium cepa* Glyox homolog 2; Accession number: AB 627990) consists of a 1240-bp nucleotide encoding a polypeptide of 364 amino acids having a predicted molecular weight of 40,329 and a predicted pI of 8.03. In order to determine whether these cDNAs encode Gly I, they were expressed as β-galactosidase fusion proteins in XL1-blue cells in the presence of 1 mM IPTG. However, we found Gly I activity only in the short-type Gly I cDNA clone. The other two types (medium- and long-) of Gly I cDNA clone had no Gly I activity. However, their presence in the onion genome might play other physiological functions which warrant further research.

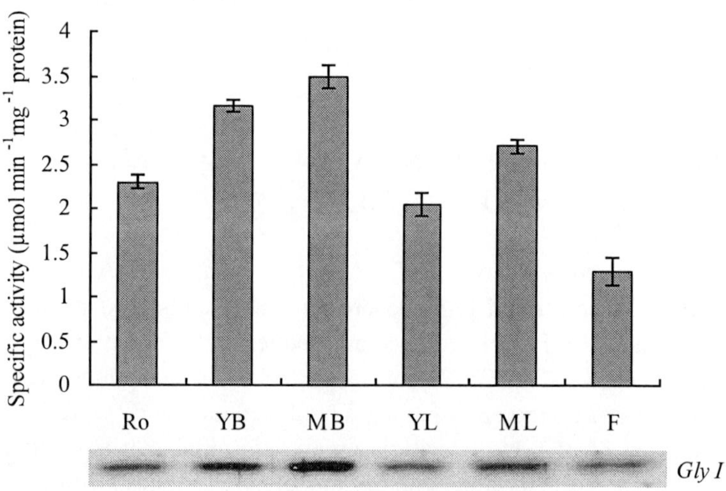

Figure 13.Western blot analysis and specific activities of Gly I in various organs of onion plants. Soluble protein was extracted from roots (Ro), young bulbs (YB), mature bulbs (MB), young leaves (YL), mature leaves (ML), and flowers (F). The specific activity in each organ was estimated and repeated three times. The bottom panel shows typical Western blot results for one experiment (adapted from Hossain & Fujita, 2009).

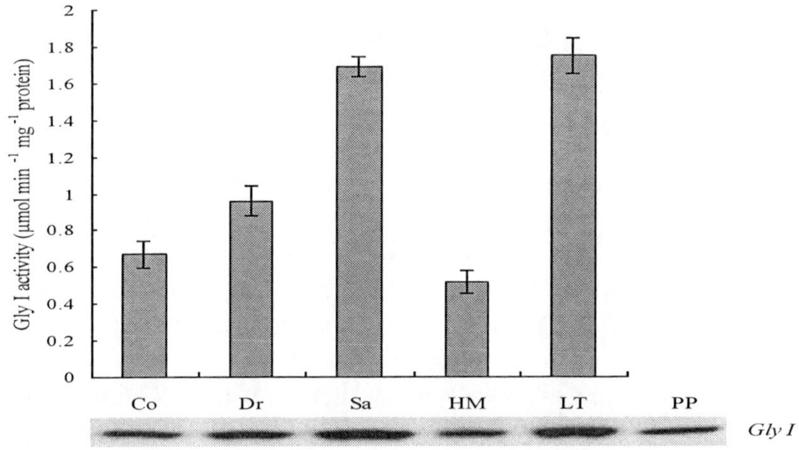

Figure 14.Effects of various stresses on Gly I activity and *Gly I* protein level. Onion callus was maintained under stressful conditions: Control (Co), Drought (Dr), salinity (Sa), heavy metal (HM), and low temperature (LT) for up to 7 d. Each lane contained 25μg of protein, except for lane 6 which contained 0.2 μg of purified Gly I protein (PP). Results were obtained from three independent experiments. The bottom panel shows typical Western blot results for one experiment (adapted from Hossain & Fujita, 2009).

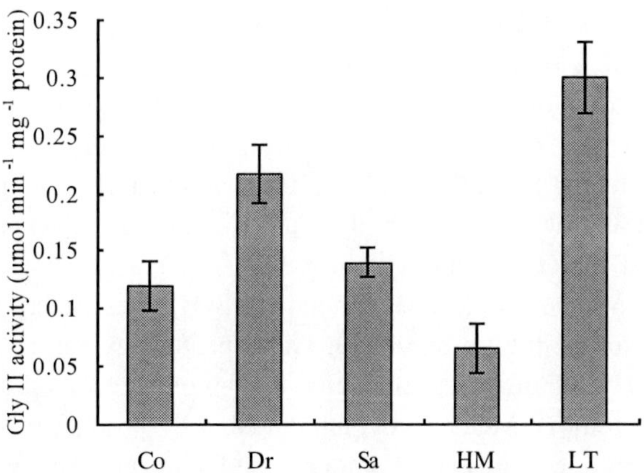

Figure 15.Effects of various stresses on Gly II activity. Onion callus was maintained under stressful conditions: Control (Co), Drought (Dr), salinity (Sa), heavy metal (HM) and low temperature (LT) for up to 7 d. Results were obtained from three independent experiments (adapted from Hossain & Fujita, 2009).

17. Induction of Glyoxalase Pathway Enzymes under Abiotic Stresses

Due to the fundamental importance of the glyoxalase pathway in plant abiotic stress response and tolerance, we studied stress-induced alteration of glyoxalase pathway enzymes in onion calli subjected to salinity (400 mM NaCl), drought (10% PEG), heavy metal (0.5 mM $CdCl_2$) and low temperature stress (4°C). A sharp increases in Gly I activity and accumulation of Gly I proteins were observed in response salinity, drought and low-temperature stress but remain unchanged in response to heavy metal stress (Figure 14). Low temperature stress (4°C) showed the highest induction (2.6-fold) of Gly I activity, followed by salinity (2.54-fold) and drought stress (1.43-fold) after 7d of treatments. A similar link of induction of Gly II activity (Figure 15) was also observed due to the various abiotic stresses, and the activity was parallel to the behaviour of Gly I, except for salinity stress. However, Gly II activity decreased in response to heavy metal stress (Figure 15). The levels of Gly I protein accumulation also showed a true reflection of possible changes in activity levels due to various abiotic stresses (Hossain & Fujita, 2009). Corroborating our results, a few recent studies also showed up-regulation of Gly I in various plant species under various abiotic stress conditions. Barley (*Hordeum vulgare* L.) genotype cv. Morex showed up-regulation of Gly I in response to salt stress (Witzel et al., 2009). Durum wheat (*Triticum durum* Desf. cv. Svevo) subjected to heat stress showed up-regulation of Gly I protein expression (Laino et al., 2010) and rice (*Oryza sativa* L. cv. Dongjin) seedlings subjected to chilling stress (10°C) showed gradual up-regulation of Gly I (Lee et al., 2009). Additionally, Tuomainen et al. (2011) did not observe an increase of the Gly I transcript in a metal hyperaccumulator plant *Thlaspi caerulescens* in response to heavy metal ($ZnSO_4$) stress. Very recently, Du et al. (2011) detected the up-regulation Gly I protein expression in rice leaves subjected toUV radiation. It is predicted that an elevated level of Gly I activity is required to remove MG, a toxic, unavoidable by-product of triosephosphate metabolism produced in ample amounts under normal and various stressful conditions (Veena et al., 1999; Yadav et al., 2005a, 2005b; El-Shabrawi et al., 2010; Hossain et al., 2009, 2011a, 2011b; Hossain & Fujita, 2009; 2010; Hoque et al., 2008; Hasanuzzaman et al., 20011a, 2011b). The simultaneous induction of both Gly I and Gly II activity in response to various stresses indicates that both enzymes are critically important in stress-induced MG detoxification and induction of stress tolerance by maintaining glutathione homeostasis (Saxena et al., 2005; Yadav et al., 2005a, 2005b). Overexpression of the glyoxalase pathway genes (Gly I and II) involved in the regulation of MG and ROS levels resulted in increased tolerance salinity and heavy metal-induced oxidative stress (Yadav et al., 2005a, 2005b; Singla-Pareek et al., 2003, 2006; 2008; Bhomker et al., 2008). It can therefore be predicated that the activation of Gly I and II genes by multistress suggests that their overexpression can lead to tolerance to multiple forms of stress in plants (Hoque et al., 2008; Hossain & Fujita, 2009, 2010, 2011; Hossain et al., 2009, 2011a, 2011b).

18. Evolutionary Relationship of the Gly I Gene

To ascertain the evolutionary relationship among the several known Gly I sequences of plants and the Gly I sequences of onion, a phylogenetic tree (Figure 16) was constructed. The tree formed two main clusters. Short-type Gly I family of both monocots and dicots grouped into one cluster whereas medium- and long-type Gly I family of both monocots and dicots (including onion Glyox homolog 1 and 2) grouped into a separate cluster. It is interesting that onion Glyox S (187) (monocot) is closely related to *Cucurbita maxima* (185)(dicot) rather than that of *Populus trichocarpa* (184) (monocot). As Gly I is a ubiquitous protein and plays an intrinsic cellular function in both higher and lower organisms, it retains its unique identity during the evolution and separation of species. Sequences of plant Gly I present in the database showed minor changes in the amino acid number of both monocot and dicot species which might not have any deleterious effects on the function and regulation of this protein. This is one of the reasonsfor forming one cluster of both monocot and dicot species. The phylogenetic analysis also showed a closer relationship of long-type Gly I of *Triticum aestivium* (284) (monocot) and *Oryza sativa* (380) (monocot) with *Arabidopsis* (283)(dicot). Importantly, *Allium*-homolog 1 (294) is closely related to *Picea sitchensis* (289)(a gymnosperm) and *Allium*-homolog 2 (364) is closely related to *Arabidopsis* (350) (a dicot species). However, sequence alignment of the deduced amino acid sequence of onion Gly I showed significant homology with other known plant Gly I sequences present in the database (Hossain & Fujita, 2009). Very high similarity was observed among the short-type Gly I sequences from start to end. In the case of medium- and long-type Gly I sequences, although there was less similarity in the initial few amino acids, higher similarity was observed in the rest of the amino acid sequences which corresponded to short-type Gly I sequences. In most cases, medium-type (L) Gly I showed high sequence similarity after 20-30 initial amino acids and a similarly long-type (LL) Gly I showed high sequences similarity after 25-30 initial amino acids except for onion Glyox LL, which showed similarity after 103amino acids, *Arabidopsis thaliana* after 89 and *Oryza sativa* LL (380) after 119. Therefore, it can be concluded that medium- and long-type Gly I evolved from the gene duplication of short-type sequences (Hossain & Fujita, 2009). A similar gene duplication event of the Gly I gene was also reported (Clugston et al., 1998). The evolutionary relationship of Gly I sequences is intriguing.

19. Conclusion

Plenty of studies on the antioxidant activity of *Allium*have been conducted due to the currentinterest in health foods. These research findings suggest that onionisa rich source of different kinds of antioxidants that couldplay important roles in physiological protection against ROS generated in the body (Aoyama &Yamamoto, 2007). Onion is a good source of quercetin, one of the most abundant flavonoids in vegetables and fruits. Onion contains other well-known antioxidant components such as volatile sulfur compounds which are known to inhibit lipid peroxidation. During the last 10 years, research with onion has significantly increased our understanding of the physiological and molecular aspects of GSH-metabolizing enzymes. Mature onion bulb express higher GSTs, Gly I and alliinase activity than other vegetable crops. Different components of GSTs (GSTa, GSTb, GSTc, GSTd and GSTe) have been purified from mature onion bulbs and partially characterized. The dominant GSTs (GSTc, GSTd and GSTe) are 27-kDa proteins but minor GSTs have higher molecular weights. We isolated and determine the structure of the physiological counterpart of onion GSTs. Although few GSH derivatives are potent inhibitors of onion GSTs, onion sulfur compounds have the least possibility of being potent physiological inhibitors of onion GSTs, which have a significant interaction with flavonoids, particularly quercetin and its precursors; however, quercetin-4′-glucoside and quercetin-3,4′-diglucoside were isolated from onion bulbs as physiological inhibitors of GSTc, GSTd and *Ac*GSTF1. The activity of GSTe varied greatly in response to growing season indicating the important biological significance of this specific GST.Molecular cloning of two GST cDNAs was performed and the genetic structure was determined. *Ac*GSTU1 is a tau type GST whereas *Ac*GSTF1 belongs to the phi type GST. Both tau and phi type GSTs are the most common, are plant-specific and are responddifferently to different stresses. Different organs of healthy onion plants and callus express GST activity and accumulate GSTs; roots, however, accumulated the highest level which is associated with plant defense against stress in the growing environment. Gly I was highly purified from onion bulbs and the specific activity of the purified enzyme is the highest reported to date in plants. The molecular weight of the purified enzyme is approximately 25,000. Highest accumulation Gly I protein was observed in mature bulbs, indicating its fundamental role in different plant physiological processes. Three types of Gly I cDNA were isolated and their genetic structure was determined. Importantly, the short-type Gly I gene and protein was found to be differentially regulated under various abiotic stress conditions. Additionally, the glyoxalase pathway enzymes (Gly I and II) and Gly I protein were responsive against various abiotic stresses indicating the roles of the glyoxalase pathway in stress tolerance. The evolutionary relationship of the Gly I gene is fascinating and it is speculated that medium- and long-type Gly I genes evolved from gene duplication of short-type Gly I.

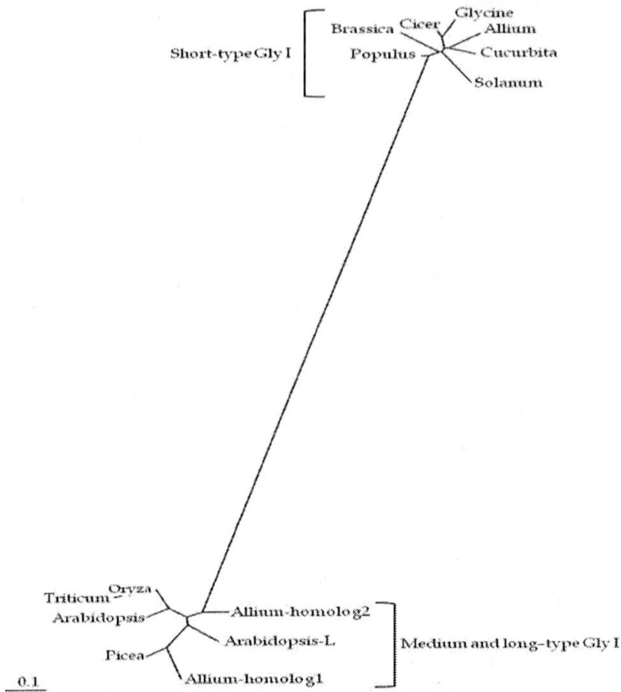

Figure 16.Phylogenetic tree showing evolutionary relationship of plant Gly I sequences.The tree was constructed using CLUSTALW software.

References

Ahmed, M.S., Ainley, K., Parish, J.H. & Hadi, S.M. (1994).Free radical-induced fragmentation of proteins by quercetin.*Carcinogenesis, 15*, 1627-1630.

Alfenito, M., Soucer, E., Goodman, C.D., Buell, R., Mol, J., Koes, R. & Walbot, V. (1998).Functional complementation of anthocyanin sequestration in the vacuole by widely divergent glutathione *S*-transferases.*Plant Cell, 10*, 1135-1149.

Anderson, J.V. & Davis, D.G. (2004). Abiotic stress alters transcript profiles and activity of glutathione *S*-transferase, glutathione peroxidase, and glutathione reductase in *Euphorbia esula. PhysiologiaPlantarum,120,* 421-433.

Aoyama,S.& Yamamoto, Y. (2007). Antioxidant activity and flavonoid content of Welsh onion (*Allium fistulosum*) andthe effect of thermal treatment.*Food Science and Technology Research, 13,* 67-72.

Arabbi, P.R., Genovese, M.I. & Lajolo, F.M. (2004). Flavonoids in vegetable foods commonly consumed in Brazil and estimated ingestion by the Brazilian people. *Journal of Agriculture and Food Chemistry,52*, 1124-1131.

Bartling, D., Radzio, R., Steiner, U. & Weiler, E.W. (1993).A glutathione *S*-transferase with glutathione- peroxidase activity from *Arabidopsis thaliana.European Journal of Biochemistry, 216*, 579-586.

Bhomkar, P., Upadhyay, C.P., Saxena, M., Muthusamy, A., Prakash, N.S., Poggin, K., Hohn, T. & Sarin, N.B. (2008). Salt stress alleviation in transgenic *Vigna mungo* L. Hepper (blackgram) by oxverexpression of the *glyoxalase I* gene using a novel *Cestrum yellow leaf curling virus* (CmYLCV) promoter.*Molecular Breeding, 22,* 169-181.

Block, E., Gulati, H., Putman, D., Sha, D., You, N. & Shu-Hai, Z. (1997). Allium chemistry: Synthesis of 1-[alk(en)ylsulfinyl]propyl alk(en)yl disulfides (cepaenes), antithrombotic flavorants from homogenates of onion (*Allium cepa*). *Journal of Agriculture and Food Chemistry,45,* 4414-4422.

Boot, K.J.M., van der Zaal, B.J., Velterop, J., Quint, A., Mennes, A.M., Hooykaas, P.J.J., Libbenga, K.R.(1993). Further characterization of expression of auxin-induced genes in tobacco (*Nicotiana tabacum*) cell-suspension cultures.*Plant Physiology,102,* 513-520.

Chakravarty, T.N. & Sopory, S.K. (1998).Blue light stimulation of cell proliferation and glyoxalase I activity in callus cultures of *Amaranthus paniculatus.Plant Sciences,132,* 63-69.

Chen, Z.Y., Brown, R.L., Damann, K.E. & Cleveland, T.E. (2004).Identification of a maize kernel stress-related protein and its effect on aflatoxin accumulation. *Phytopathology*, *94,* 938-945.

Clugston, S.L., Daub, E. & Honek, J.F. (1998). Identification of glyoxalase I sequences in *Brasssica oleracea* and *Sporobolus stapfinus*: evidence for gene duplication events. *Journal of Molecular Evolution,47,* 230-234.

Cummins, I., Cole, D.J. & Edwards, R. (1997). Purification of multiple glutathione transferases involved in herbicide detoxification from wheat (*Triticum aestivum* L.) treated with the safener fenchlorazole-ethyl. *Pesticide Biochemistry and Physiology,59,* 35-49.

Cummins, I., O'Hagen, D., Jablonkai, I., Cole, D.J., Hehn, A. & Werckreichhart, D. (2003). Cloning, characterization and regulation of a family of phi class glutathione transferases from wheat. *Plant Molecular Biology,52,* 591-603.

Deswal, R., Chakravarty, T.N. & Sopory, S.K., (1993). The glyoxalase system in higher plants: regulation in growth and differentiation. *Biochemical Society Transactions*, *21,*527-530.

Deswal, R. & Sopory, S.K. (1991). Purification and partial characterization of glyoxalase I from a higher plant *Brassica juncea.FEBS Letters*, *282,* 277-280.

Deswal, R., & Sopory, S.K. (1998). Biochemical and immunochemical characterization of *Brassica juncea* glyoxalase I. *Phytochemistry*, *49,* 2245-2253.

Diao, G., Wang, Y., Wang, C. & Yang, C. (2011).Cloning and functional characterization of a novel glutathione *S*-transferase gene from *Limonium bicolor.Plant Molecular Biology Reporter*, *29,* 77-87.

Dixon, D.P. & Edwards, R. (2010).Glutathione transferases. In:*The Arabidopsis Book,* 8:e0131.

Dixon, D.P., Cole, D.J. & Edwards, R. (1998).Purification, regulation and cloning of a glutathione transferase (GST) from maize resembling the auxin-inducible type-III GSTs. *Plant Molecular Biology,36,* 75-87.

Dixon, D.P., Davis, B.G. & Edwards, R. (2002).Functional divergence in the glutathione transferase superfamily in plants- identification of two classes with putative functions in redox homeostasis in *Arabidopsis thaliana.Journal of Biological Chemistry, 277*, 30859-30869.

Dorant, E., Van Din Brandt, P. & Goldbohm, R. (1994).A prospective cohort study on *Allium* vegetable consumption, garlic supplement use, and the risk of lung carcinoma in the Netherlands.*Cancer Research,54*, 6148-6153.

Dorsch, W. & Wagner, H. (1991).New antiasthmatic drugs from traditional medicine?*International Archives of Allergy and Applied Immunology*, *94*, 262-265.

Du, H., Liang, Y., Pei, K. & Ma, K. (2011). UV radiation-responsive proteins in rice leaves: a proteomic analysis. *Plant and Cell Physiology*, *52*, 306-316.

Edwards, R. & Dixon, D.P. (1991).Glutathione *S*-cinnamoyl transferases in plants.*Phytochemistry, 30*, 79-84.

Edwards, R., Dixon, D.P. & Walbot, V. (2000). Plant glutathione *S*-transferases: enzymes with multiple functions in sickness and in health. *Trends in Plant Science,5*, 193-198.

El-Shabrawi, H., Kumar, B., Kaul, T., Reddy, M.K., Singla-Pareek, S.L. & Sopory, S.K. (2010).Redox homeostasis, antioxidant defense, and methylglyoxal detoxification as markers for salt tolerance in Pokkali rice. *Protoplasma*, *245*, 85-96.

Espartero, J., Aguayo, I.S.& Pardo, J.M. (1995).Molecular characterization of glyoxalase-I from a higher plant; upregulation by stress. *Plant Molecular Biology*, *29*, 1223-1233.

Ezaki, B., Suzuki, M., Motoda, H., Kawamura, M., Nakashima, S. & Matsumoto, H. (2004). Glutathione *S*-transferase, *AtGST1*, and *AtGST11* in response to aluminum stress. *Plant Physiology*, *134*, 1-11.

Fatima, R.A. & Ahmed, M. (2005).Certain antioxidant enzymes of *Alliun cepa* as biomarkers for the detection of toxic heavy metals in wastewater.*Science of the Total Environment,346*, 256-273.

Fitzpatrick, D., Hirschfield, S. & Coeffey, R. (1993).Endothelium-dependent vasorelaxing activity of wine and other grape products.*The American Physiological Society, 265*,774-778.

Fossen, T., Pedersen, A.T. & Andersen, Ø.M. (1998).Flavonoids from red onion (*Allium cepa*).*Phytochemistry*, *47*, 281-285.

Foyer, C.H. & Noctor, G. (2011). Ascorbate and glutathione: the heart of the redox hub. *Plant Physiology*, *155*, 2-18.

Fujita, M. & Hossain, M.Z. (2003a).Molecular cloning of cDNA for three tau-type glutathione *S*-transferases in pumpkin (*Cucurbita maxima*) and their expression properties.*Physiologia Plantarum*, *117*,85-92.

Fujita, M. & Hossain, M.Z. (2003b).Modulation of pumpkin glutathione *S*-transferases by aldehydes and related compounds. *Plant and Cell Physiology*,*44*,481-490.

Fujita, M., Adachi, Y. & Hanada, Y. (1994). Preliminary characterization of glutathione *S*-transferases that accumulate in callus cells of pumpkin (*Cucurbita maxima* Duch.). *Plant and Cell Physiology,35*, 275-282.

Fujita, M., Adachi, Y. & Sakato, N. (1998). Purification of pumpkin glutathione *S*-transferase species specially present in cultured cells treated by excessive concentrations of 2,4-dichlorophenoxyacetic acid but absent in normal plants. *Bioscience, Biotechnology, and Biochemistry, 62*, 2431-2434.

Galle, Á., Csiszár, J., Secenji, M., Tari, I., Guóth, A., Györgyey, J. & Erdei, L. (2008). Monitoring the levels of phi and tau group GST genes in wheat cultivars under osmotic stress.*Acta Biologica Szegediensis*, *52*, 95-96.

Gazzani, G., Papetti, A., Massolini, G. & Daglia, M. (1998). Anti- and prooxidant activity of water soluble components of some common diet vegetables and the effect of thermal treatment.*Journal of Agriculture and Food Chemistry,46*, 4118-4122.

Goldman, I., Kopelberg, M., Devaene, J. & Schwartz, B. (1996). Antiplatelet activity in onion is sulfur dependent. *Thrombosis and Haemostasis*, *76*, 450-452.

Gong, H., Jiao, Y., Hu, W. & Pua E-C.(2005). Expression of glutathione *S*-transferase and its role in plant growth and development *in vivo*and shoot morphogenesis *in vitro*.*Plant Molecular Biology*, *57*, 53-66.

Gorinstein, S., Leontowicz, H., Leontowicz, M., Namiesnik, J., Najman, K., Drzewiecki, J., Cvikrová, M., Martincová,O., Katrich, E. & Trakhtenberg, S. (2008).Comparison of the main bioactive compounds and antioxidant activities in garlic and white and red onions after treatment protocols.*Journal of Agriculture and Food Chemistry*, *56*, 4418-4426.

Hahn, K. & Strittmater, G. (1994).Pathogen-defence gene prp1-1 from potato encodes an auxin-responsive glutathione *S*-transferase. *European Journal of Biochemistry, FEBS,226*, 619-626.

Halušková,L., Valentovičová,K., Huttová,J., Mistrík, I. & Tamás,L. (2009).Effect of abiotic stresses on glutathione peroxidase and glutathione *S*-transferase activity in barley root tips.*Plant Physiology and Biochemistry*, *47*, 1069-1074.

Harborne,J.B., Heller,W. &Forkmann,G. *(*1993*)*.Biosynthesis.*In*Harborne,J.B. (ed.), *The Flavonoids. Advances in Research since 1986*, *pp*499-535, London, Chapman and Hall Ltd.

Hasanuzzaman, M., Hossain M.A. & Fujita, M. (2011a).Selenium-induced up-regulation of the antioxidant defense and methylglyoxal detoxification system reduces salinity-induced damage in rapeseed seedlings. *Biological Trace Element Research*, DOI 10.1007/s12011-011-8958-4.

Hasanuzzaman, M., Hossain, M.A. & Fujita, M. (2011b). Nitric oxide modulates antioxidant defense and the methylglyoxal detoxification system and reduces salinity-induced damage of wheat seedlings. *Plant Biotechnology Reports*, (in press).

Hertog, M.G.L., Feskens,E.J.M., Hollman,P.C.H., Katan,M.B. &Kromhout,D. (1993).Dietary antioxidant flavonoids and risk of coronary heart disease: The Zutphen elderly study. *Lancet, 342*, 1007-1011.

Hoque, M.A., Banu, M.N.A., Nakamura, Y., Shimoishi, Y. & Murata, Y. (2008). Proline and glycinebetaine enhance antioxidant defense and methylglyoxal detoxification systems and reduce NaCl-induced damage in cultured tobacco cells. *Journal of Plant Physiology*, *165*, 813-824.

Hoque, M.A., Uraji, M., Banu, M.N.A., Mori, I.C., Nakamura, Y. & Murata Y. (2010).The effect of methylglyoxal on glutathione *S*-transferase from *Nicotiana tabacum.Bioscience, Biotechnology, and Biochemistry*, *74*, 2124-2126.

Hossain, M.A. &Fujita, M. (2009).Purification of glyoxalase I from onion bulbs and molecular cloning of its cDNA.*Bioscience, Biotechnology, and Biochemistry*, *73*, 2007-2013.

Hossain, M.A. & Fujita, M. (2010).Evidence for a role of exogenous glycinebetaine and proline in antioxidant defense and methylglyoxal detoxification systems in mung bean seedlings under salt stress.*Physiology and Molecular Biology of Plants*, *16*, 19-29.

Hossain, M.A. & Fujita, M. (2011).Regulatory role of components of ascorbate-glutathione (AsA-GSH) pathway in plant tolerance to oxidative stress. In: Anjum, N.A., Umar, S. & Ahmed, A. (eds.), *Oxidative Stress in Plants: Causes, Consequences and Tolerance*, pp 81-147, India, IK International Publishing House Pvt. Ltd.

Hossain, M.A., Hasanuzzaman, M. & Fujita, M. (2010). Up-regulation of antioxidant and glyoxalase systems by exogenous glycinebetaine and proline in mung bean confer tolerance to cadmium stress. *Physiology and Molecular Biology of Plants*, *16*, 259-272.

Hossain, M.A., Hasanuzzaman, M.& Fujita, M. (2011a).Coordinate induction of antioxidant defense and glyoxalase system by exogenous proline and glycinebetaine is correlated with salt tolerance in mung bean. *Frontiers of Agriculture in China*, *5*, 1-14.

Hossain, M.A., Hossain, M.Z. & Fujita, M. (2009).Stress-induced changes of methylglyoxal level and glyoxalase I activity in pumpkin seedlings and cDNA cloning of glyoxalase I gene. *Australian Journal of Crop Science*, *3*, 53-64.

Hossain, M.A., Teixeira da Silva, J.A. & Fujita, M. (2011b). Glyoxalase system and reactive oxygen species detoxification system in plant abiotic stress response and tolerance: An intimate relationship. In:Shanker, A.K. &Venkateswarlu, B. (eds.), *Abiotic Stress in Plants-Mechanisms and Adaptations*, pp. 235-266, Croatia, INTECH-Open Access Publisher.

Hossain, M.D., Rohman, M.M. & Fujita, M. (2007a). Comparative investigation of glutathione *S*-transferases, glyoxalase-I and alliinase activities in different vegetable crops.*Journal of Crop Science and Biotechnology,10*, 21-28.

Hossain, M.D., Rohman, M.M. & Fujita, M. (2008). Inhibitors of plant glutathione *S*-transferases: their physiological counterparts in pumpkin and onion plants. *Current Topics in Plant Biology, 9*, 27-43.

Hossain, M.D., Suzuki, T. & Fujita, M. (2007b).A preliminary approach to identify the physiological substrates of pumpkin glutathione *S*-transferase through inhibition studies.*Acta Horticulturae*, *731*, 217-222.

Hossain, M.Z. & Fujita, M. (2002).Purification of a phi-type glutathione *S*-transferase from pumpkin flowers and molecular cloning of the cDNA.*Bioscience, Biotechnology, and Biochemistry*, *66*, 2068-2076.

Hossain, M.Z., Hossain, M.D. & Fujita, M. (2006a). Induction of pumpkin glutathione *S*-transferase by different stresses and its possible mechanisms.*Biologia Plantarum*, *50*, 210-218.

Hossain, M.Z., Teixeira da Silva, J.A. & Fujita, M. (2006b). Differential roles of glutathione *S*-transferase in oxidative stress modulation.In:Teixeira da Silva, J.A. (ed.), *Floriculture, Ornamental and Plant Biotechnology. Advances and Topical Issues*, Vol. 3, pp. 108-116, UK, Global Science Books Ltd.

Hu, Y., Ge, Y., Zhang, C., Ju, T. & Cheng, W. (2009). Cadmium toxicity and translocation in rice seedlings are reduced by hydrogen peroxide pretreatment. *Plant Growth Regulation*, *59*, 51-61

Jain, M., Choudhary, D., Kale, R.K. & Bhalla-Sarin, N. (2002). Salt and glyphosphate-induced increase in glyoxalase I activity in cell lines of groundnut (*Arachis hypogaea*). *Physiologia Plantarum, 114*,499-505.

Jain, M., Ghanashyam, C. & Bhattacharjee, A. (2010). Comprehensive expression analysis suggests overlapping and specific roles of rice glutathione *S*-transferase genes during development and stress responses. *BMC Genomics*, *11*, 73.

Jha B., Sharma A. & Mishra, A. (2010). Expression of *SbGSTU* (*tau* class glutathione *S*-transferase) gene isolated from *Salicornia brachiata* in tobacco for salt tolerance. *Molecular Biology Reports*, DOI:10.1007/s11033-010-0625-x

Jones, M.G., Hughes, J., Tregova, A., Milne, J., Tomsett, A.B. & Collin, H.A. (2004).Biosynthesis of the flavour precursors of onion and garlic.*Journal of Experimental Botany,55*, 1903-1918.

Kaur, C., Joshi, S. & Kapoor, H.C. (2007).Antioxidants in onion (*Allium cepa* L.) cultivars grown in India.*Journal of Food Biochemistry*, *33*,184-200.

Kumar, V. & Yadav, S.K. (2009). Proline and betaine provide protection to antioxidant and methylglyoxal detoxification systems during cold stress and *Camellia sinensis* (L.) O.Kuntze. *Acta Physiologia Plantarum,31*, 261-269.

Lachman, J., Pronek, D., Hejtmankova, A., Dudjak, J., Pivec, V. & Faitova, K. (2003). Total polyphenol and main flavonoid antioxidants in different onion (*Allium cepa* L.) varieties.*Horticulture Science(Prague),30*, 142-147.

Laino, P., Shelton, D., Finnie, C., De Leonardis, A.M., Mastrangelo, A.M., Svensson, B., Lafiandra, D. & Masci, S. (2010). Comparative proteome analysis of metabolic proteins from seeds of durum wheat (cv. Svevo) subjected to heat stress. *Proteomics*, *10*, 2359-2368.

Lamoureux, G.L. & Rusness, D.G. (1980).In vitro metabolism of pentachloronitrobenzene to pentachloromethylthiobenzene by onion: Characterization of glutathione *S*-transferase, cysteine CS lyase, and *S*-adenosylmethionine methyl transferase activities. *Pesticide Biochemistry and Physiology,14*, 50-61.

Lancaster, J.E. & Boland, M.J. (1990).Flavor biochemistry. In: Rabinowitch, H. & Brewster, J. (eds,), *Alliums and Allied Crops*, Vol 3. pp 33-72, Boca Raton, FL: CRC Press.

Lancaster, J.E. & Kelly, K.E. (1983). Quantitative analysis of the *S*-alk(en)yl-L-cysteine sulphoxides in onion (*Allium cepa* L.). *Journal of Science in Food Agriculture,34*, 1229-1235.

Lancaster, J.E., Shaw, M.L., Joyce, M.D.P., McCallum, J.A. & McManus, M.T. (2000).A novel alliinase from onion roots.Biochemical characterization and cDNA cloning.*Plant Physiology,* 122, 1269-1279.

Lee, D.G., Ahsan, N., Lee, S.H., Lee, J.J., Bank, J.D., Kang, K.Y. & Lee, B.H. (2009). Chilling stress-induced proteomic changes in rice roots. *Journal of Plant Physiology*, *166*, 1-11.

Lin, F., Xu, J., Shi, J., Li, H. & Li, B. (2010).Molecular cloning and characterization of a novel glyoxalase I gene *TaGly I* in wheat (*Triticum aestivum* L.).*Molecular Biology Reports*, *37*, 729–735.

Lopez, M.F., Patton, W.F., Sawlivich, W.B., Erdjument-Bromage, H. & Barry, P. (1994). A glutathione *S*- transferase (GST) isozyme from broccoli with significant sequence homology to the mammalian theta-class of GSTs. *Biochimica et Biophysica Acta,1205*, 29-38.

Malaveille, C., Hautefeuille, A., Pignatelli, B., Talaska, G., Vineis, P. & Bartsch, H. (1996).Dietary phenolics as anti-mutagens and inhibitory of tobacco related DNA adduction in the urothelium of smokers.*Carcinogenesis*, *17*, 2193-2200.

Mang, H.G., Kang, E.O., Shim, J.H., Kim, S., Park, K.Y., Kim, Y.S., Bahk, Y.Y. & Kim, W.T. (2004). A proteomic analysis identifies glutathione *S*-transferase isoforms whose abundance is differentially regulated by ethylene during the formation of early root epidermis in Arabidopsis seedlings. *Biochimica et Biophysica Acta*, *1676*, 231-239.

Mannervik, B. & Danielson, U.H. (1988).Glutathione transferases - structure and catalytic activity.*CRC Critical Reviews in Biochemistry*,*23*, 283-337.

Marrs, K.A. & Walbot, V. (1997). Expression and RNA splicing of the maize glutathione *S*-transferase Bronze2 gene is regulated by cadmium and other stresses. *Plant Physiology,113*, 93-102.

Marrs, K.A. (1996). The functions and regulation of glutathione *S*-transferases in plants.*Annual Review of Plant Physiology, Plant Molecular Biology,47*, 127-158.

Martins, A.M.T.B.S., Cordeiro, C.A.A. & Freire, A.M.J.P. (2001).*In situ* analysis of methylglyoxal metabolism in *Saccharomyces cerevisiae*.*FEBS Letters,499*, 41-44.

McClure,J.W. (1986). Physiology of flavonoids in plants. *In*Cody,V., Middleton,E. &Harborne,J.B. (eds.),*Plant Flavonoids in Biology and Medicine: Biochemical, Pharmacological, and Structure-Activity Relationships, pp*77-85, New York, Alan R. Liss, Inc.

McGonigle, B., Keeler, S.J., Lau, S.M.C., Koeppe, M.K. & O'Keefe, D.P. (2000).A genomic approach to the comprehensive analysis of the glutathione *S*-transferase gene family in soybean and maize.*Plant Physiology, 124*, 1105-1120.

Middleton,E. &Kandaswami,C. *(*1992*)*.Effects of flavonoids on immune and inflammatory cell function. *Biochemical Pharmacology,43*, 1167-1179.

Mittler, R., Vanderauwera, S., Gollery, M. & Breusegem, F.V. (2004).Reactive oxygen gene network of plants.*Trends in Plant Science, 9*,490-498.

Moskaug, J.O., Carlsen, H., Myhrstad, M. & Blomhoff, R. (2004).Molecular imaging of the biological effects of quercetin and quercetin rich foods.*Mechanisms of Ageing and Development*, *125*,315-324.

Mueller, L.A., Goodman, C.D., Silady, R.A. & Walbot, V. (2000). AN9, a petunia glutathione *S*-transferase required for anthocyanin sequestration, is a flavonoid-binding protein. *Plant Physiology*, *123*, 1561-1570.

Nakagawa, K., Ikeuchi, M. & Tsugita, Y. (1986).Glutathione content of vegetables.*Journal of Food Hygiene Society, Japan, 27*, 425-427.

Nock, L.P. & Mazelis, M. (1987). The C-S lyases of higher plants: direct comparison of the physical properties of homogeneous alliin lyase of garlic (*Allium sativum*) and onion (*Allium cepa*). *Plant Physiology*, *85*, 1079-1083.

Nock, L.P. & Mazelis, M. (1989).Lack of homology between the alliin lyases of garlic and onion.*Phytochemistry*, *28*, 729-731.

Paulus, C., Kollner, B. & Jacobsen, H.J. (1993). Physiological and biochemical characterization of glyoxalase I: a general marker for cell proliferation from a soybean cell suspension. *Planta*, *189*, 561-566.

Polidoros, A.N. & Scandalios, J.G. (1999).Role of hydrogen peroxide and different classes of antioxidants in the regulation of catalase and glutathione *S*-transferase gene expression in maize (*Zea mays*L.).*Physiologia Plantarum*, *106*, 112-120.

Racker, E. (1951).The mechanism of action of glyoxalase.*Journal of Biological Chemistry*, 190, 685-696.

Rhodes, M. & Price, K. (1996).Analytical problems in the study of flavonoid compounds in onions.*Food Chemistry,57*, 113-117.

Rodrigues, A.S. Pérez-Gregoriob, M.R. García-Falcón M.S. & Simal-Gándara, J. (2009).Effect of curing and cooking on flavonols and anthocyanins in traditional varieties of onion bulbs.*Food Research International,42*, 1331-1336.

Rohman, M.M., Hossain, M.A. & Fujita, M. (2010b). Levels of a dominant Glutathione *S*-transferase in onion bulbs have a seasonal relationship with physiological inhibitors. *Australian Journal of Crop Science*, *4*, 55-62.

Rohman, M.M.,Hossain, M.D., Suzuki, T., Takada, G. & Fujita, M. (2009a). Quercetin-4′-glucoside: a physiological inhibitor of the activities of dominant glutathione *S*-transferases in onion (*Allium cepa* L.) bulb. *Acta Physiologia Plantarum*, *31*, 301-309.

Rohman, M.M., Suzuki, T. & Fujita, M. (2009b).Identification of a glutathione *S*-transferase inhibitor in onion bulb (*Allium cepa* L.).*Australian Journal of Crop Science*, *3*, 28-36.

Rohman, M.M., Uddin, M.S. & Fujita, M. (2010a). Up-regulation of onion bulb glutathione *S*-transferases (GSTs) by abiotic stresses: A comparative study between two differently sensitive GSTs to their physiological inhibitors. *Plant Omics Journal*, *3*, 28-34.

Roldán-Marín,E., Sánchez-Moreno,C., Lloría,R., de Ancos,B.& Cano,M.P. (2009). Onion high-pressure processing: Flavonol content and antioxidant activity. *Food Science and Technology*, *42*, 835-841.

Roxas, V.P., Smith, R.K., Allen, E.R. & Allen, R.D. (1997). Overexpression of glutathione *S*-transferase glutathione peroxidase enhances the growth of transgenic tobacco seedlings during stress. *Nature Biotechnology*, *15*, 988-991.

Saxena, M., Bisht, R., Roy, D.S., Sopory, S.K. & Bhalla-Sarinn, M. (2005).Cloning and characterization of a mitochondrial glyoxalase II from *Brassica juncea* that is upregulated by NaCl, Zn and ABA.*Biochemical and Biophysical Research Communication,336*, 813-819.

Schröder, P. & Stampfl, A. (1999).Visualization of glutathione conjugation and induction of glutathione *S*-transferases in onion (*Allium cepa*L.) epidermal tissue.*Zeitschrift für Naturforschung,54C*, 1033-1041.

Singla-Pareek, S.L., Ray, M., Reddy, M.K. & Sopory, S.K. (2003).Genetic engineering of the glyoxalase pathway in tobacco leads to enhanced salinity tolerance.*Proceedings of the National Academy of Sciences of USA,100*, 14672-14677.

Singla-Pareek, S.L., Yadav, S.K., Pareek, A., Reddy, M.K. & Sopory, S.K. (2006). Transgenic tobacco overexpressing glyoxalase pathway enzymes grow and set viable seeds in zinc-spiked soils. *Plant Physiology*, *140*, 613-623.

Singla-Pareek, S.L.,Yadav, S.K., Pareek, A.,Reddy, M.K. & Sopory, S.K. (2008).Enhancing salt tolerance in a crop plant by overexpression of glyoxalase II. *Transgenic Research*, *17*, 171-180.

Skipsey, M., Andrews, C.J., Townson, J.K., Jepson, I. & Edwards, R. (2000).Cloning and Characterization of Glyoxalase I from Soybean.*Archives of Biochemistry and Biophysics*, *374*, 261-268.

Sobhanian, H., Motamed, N., Jazil, F.R., Nakamura, T. & Komatsu, S. (2010). Salt stress induced differential proteomic and metabolome response in the shoots of *Seluropus lagopoides* (Poaceae), a halophyte C_4 plant. *Journal of Proteomic Research*, 9, 2882-2897

Szent-Gyorgyi, A., Egyud, L.G. & McLaughlin, J.A. (1967).Ketoaldehydes and cell division.*Science*, *155*, 539-541.

Tailor, L.P. & Grotewold, E. (2005). Flavonoids as developmental regulators.*Current Opinion in Plant Biology,8*, 317-323.

Tabrez, S. & Ahmad, M. (2011). Components of antioxidative system in *Allium cepa*as the toxicity monitor of trichloroethylene (TCE). *Toxicological and Environmental Chemistry*, *93*, 73-84.

Thornalley, P.J. (1990). The glyoxalase system: new developments towards functional characterization of a metabolic pathway fundamental to biological life. *Biochemical Journal,269*, 1-11.

Thornalley, P.J. (1996). Pharmacology of methylglyoxal: formation, modification of proteins and nucleic acids, and enzymatic detoxification-a role in pathogenesis and antiproliferative chemotherapy. *General Pharmacology*, *27*, 565-573.

Tobkin, H.E. & Mazelis, M. (1979). Alliin lyase: preparation and characterization of the homogenous enzyme from onion bulbs. *Archives of Biochemistry and Biophysics*, *193*, 150-157.

Tuomainen, M., Ahonen,V., Karenlampi,S.O., Schat, H., Paasela, T.,Svanys, A., Tuohimetsa, S.,Peraniemi, S. Tervahauta, A. (2011). Characterization of the glyoxalase 1 gene *TcGLX1* in the metal hyperaccumulator plant *Thlaspi caerulescens*. *Planta*, *233*, 1173-1184.

van der Kop, D.A.M., Schuyer, M., Scheres, B., van der Zaal, B.J. & Hooykaas, P.J.J. (1996). Isolation and characterization of an auxin-inducible glutathione *S*-transferase gene of *Arabidopsis thaliana.Plant Molecular Biology,30*, 839-844.

Veena, Reddy, V.S. & Sopory, S.K. (1999). Glyoxalase I from *Brassica juncea*: molecular cloning, regulation and its over-expression confer tolerance in transgenic tobacco under stress. *The Plant Journal*, *17*, 385-395.

Walbot, V., Mueller, L. Silady, R.A. & Goodman, C.D. (2000). Do glutathione *S*-transferases act as enzymes or as carrier proteins for their natural substrates? In Brunold, C. Rennenberg, H. De Kok, L.J., Stulen I. &. Davidian. J.-C (eds.), Sulfur Metabolism in Higher Plants - Molecular, Biochemical and Physiological Aspects, pp. 155-165. Berne: Paul Haupt Publisher.

Wang, J., Cao, Y., Sun, B., Wang, C.& Mo, Y. (2011a).Effect of ultrasound on the activity of alliinase from fresh garlic.*Ultrasonics Sonochemistry*,*18*, 534–540.

Wang, Y., Qian, Y., Hu, H., Xu, Y. & Zhang, H. (2011b).Comparative proteomic analysis of Cd-responsive proteins in wheat roots.*Acta Physiologia Plantarum*, *33*,349-357.

Witzel, K., Weidner, A., Surabhi, G.K., Börner, A. & Mock, H.P. (2009).Salt stress-induced alterations in the root proteome of barley genotypes with contrasting response towards salinity.*Journal of Experimental Botany*, *60*, 3545-3557.

Yadav, S.K., Singla-Pareek, S.L., Ray, M., Reddy, M.K. & Sopory, S.K. (2005a). Methylglyoxal levels in plants under salinity stress are dependent on glyoxalase I and glutathione. *Biochemical and Biophysical Research Communication*, *337*, 61-67.

Yadav, S.K., Singla-Pareek, S.L., Ray, M., Reddy, M.K. & Sopory, S.K. (2005b). Transgenic tobacco plants overexpressing glyoxalase enzymes resist an increase in methylglyoxal and maintain higher reduced glutathione levels under salinity stress. *FEBS Letters*, *579*, 6265-6271.

Zhang, X. & Komatsu, S. (2007). A proteomic approach for identifying osmotic-stress-related proteins in rice.*Phytochemistry*, *68*,426-437.

Zhou, J. & Goldsbrough, P.B. (1993). An Arabidopsis gene with homology to glutathione *S*-transferase is regulated by ethylene. *Plant Molecular Biology*, *22*,517-523.

In: Onion Consumption and Health
Editors: C. B. Aguirre et al.
ISBN: 978-1-62100-836-1

Chapter 3

ONION CONSUMPTION AND HEALTH

Francisco Segovia and María Pilar Almajano*

ABSTRACT

Since ancient times, onions, *Allium cepa* L., one of the most widely distributed vegetables, have been used as common foods and for the treatment of many diseases. Onion is a food widely used in gastronomy and, also, for medicinal applications. It is one of the most cultivated and consumed vegetables on the planet (about 66 million tons in 2008) and one of the main ingredients of the Mediterranean diet. This plant is a rich source of several phytonutrients recognized as important elements of the Mediterranean diet and it is known as a good source of bioactive compounds, such as sulphur-containing compounds and flavonoids. Carbohydrates in onion constitute about 80% of the dry matter, and the major non-structural carbohydrates of the onion bulb are fructo-oligosaccharides, well known as fructan, followed by glucose, fructose and sucrose.. Various protein sources can be found in intact onion in low concentrations, including lectins (the most abundant proteins in onion), prostaglandins, … Also, can be found vitamins (B1, B2, B6, C, E, biotin and nicotinic acid), fatty acids, glycolipids, phospholipids and essential amino acids. But, the most recent attention has focused on the flavonoids that are also responsible for a great part of the health benefits. It is known that onions are a rich source of dietary flavonoids with average content ranging from 270 to 1917 mg of flavonoids per Kg of fresh weight,

* Email,: m.pilar.almajano@upc.edu

depending on the onion variety. They are mainly represented by the flavonols quercetin and kaempferol and other flavonoids such as isorhamnetin, myricetin and their conjugates (as glycosylated forms). Anthocyanins, namely peonidin 3-glucoside, cyanidin 3-glucoside and cyanidin 3-arabinoside and their malonylated derivatives, cyanidin 3-laminariobioside and delphinidin and petunidin derivatives, are located in the skin of red onions and in the outer fleshy layer. Previous studies indicate that flavonoids have beneficial health effects due to their antioxidant, anti-inflammatory, cardioprotective, and anticancer effects, which are of particular interest in the prevention of chronic diseases, including cardiovascular disease and some cancers. Onion extract has been recently reported to be effective in reducing risk of cardiovascular disease because of its hypocholesterolemic, hypolipidemic, anti-hypertensive, antithrombotic and anti-hyperhomocysteinemic effects, and to possess many other biological activities including antimutagenic, antiasthmatic, immunomodulatory and prebiotic activities. For this reason can be used as alternative to additive because nowadays consumers are demanding minimally processed foods without synthetic preservatives, due mainly to concerns about possible adverse health effects.

It should be noted that flavonoids of the onion are more stable than the other components. Traditionally, it was suggested that the sulfur compounds are those that are good for health because they are responsible for the characteristic flavour, aroma and tear promoting effects of the plant. These substances, highly volatile and unstable, are released when the onion is damaged or cut. The most important organosulfur compounds are the cysteine sulfoxides and the gamma-glutamylcysteines.

Onions and onion by-products have been proposed as food ingredients which can be added to food preparations to enhance their preservation during their shelf-life.

Onion is widely accepted as a good alternative as a food preservative, especially in fresh or minimally processed foods. The organosulfur compounds have been shown prevent formation of reactive oxygen species and reduction of cholesterol levels.

Epidemiological studies have shown an inverse relationship between consumption of vegetables and the risk of cardiovascular disease. The beneficial effect of flavonoids in protection against cardiovascular disease was demonstrated. By conclusion, the onions are very suitable as an ingredient in food preparation or for direct consumption. There has been increased consumption of this vegetable because of a greater public interest in natural remedies and non-conventional therapies for the treatment of certain diseases. Many consumers believe that these therapies have fewer side effects than pharmaceutical products. In spite of numerous studies about the beneficial effects of onions on human health, certain aspects still need to be investigated. Onion could be added like functional ingredients into ”fast foods”, which are widely consumed in current society, to provide them with antioxidant compounds, such as quercetin, prebiotics, or mineral nutrients, such as selenium, to prevent nutritional deficiencies.

INTRODUCTION

Allium cepa L or bulb onion, is a biennial herbaceous plant family Amaryllidaceae. These plants produce a large bulb in the first year of growth. Botanically, the Alliums are included in the *Liliaceae*.

The bulb onion is thought to have originated in central Asia (Hanelt, 1990) but is not known in the wild, so it may have become extinct or be a hybrid from ancient cultivation. Related cultivated species include garlic, leek, chives, bunching onion and some other species of significant local importance. They are generally adapted to temperate environments with low or erratic supplies of water, either in areas of sparse vegetation or accompanying grasses and wild species are found across a wide range of latitudes and altitudes in Europe, Asia, N. America and Africa (Cho, 2010).

Since ancient times, onions, *Allium cepa L.*, have been used as common foods and for the treatment of many diseases. The use of onion can be traced back to the Egyptians. The first citation of the medical use of onion is found in the Codex Ebers (1550 BC), an Egyptian medical papyrus reporting several therapeutic formulas based on garlic and onions as useful remedies for a variety of diseases such as heart problems, headache, bites, worms and tumours (Lanzotti, 2006; Slimestad, 2007). Moreover, olympic athletes were fed onions and garlic to improve performance in track and field events and europeans have treated blood clots in horses for centuries with onion. These “*folklore traditions*” thus suggested an involvement in improved blood circulation.

Also, onion extract has been recently reported to be effective in reducing risk of cardiovascular disease because of its hypocholesterolemic, hypolipidemic, anti-hypertensive, antithrombotic and anti-hyperhomocysteinemic effects, and to possess many other biological activities including antimicrobial, antioxidant, anticarcinogenic, antimutagenic, antiasthmatic, immunomodulatory and prebiotic activities (Corzo-Martinez, 2007).

In this chapter it will be attempt to analyze scientifically proven aspects, starting with the composition, metabolism and antioxidant and antimicrobial activities

Onion Composition

The organosulfur compounds in alliums have been widely studied. The most important organosulfur compounds are the cysteine sulfoxides and the gamma-glutamylcysteines. When the tissues of alliums are disrupted, the enzyme allinase hydrolyses the flavour precursors S-alk(en)yl-L-cysteine sulfoxides (ACSOs), giving rise to the flavour and pungency characteristic of the allium plants. There are four different ACSOs identified in allium species: (+)-S-allyl-L-cysteine sulfoxide known as alliin (2-PCSO), (+)-S-2-methyl-Lcysteine sulfoxide or methiin (MCSO), S-propyl-L-cysteine sulfoxide or propiin (PCSO) and (+)-S-trans-1-propenyl-L-cysteine sulfoxide or isoalliin (TPCSO) (Rose et al., 2005).

The characteristic flavour of each allium species depends on the concentrations and structures of these ACSOs. For example, isoalliin is the major sulfoxide present in onion tissues, and its hydrolysis products contribute to the flavour and lachrymatory effect (Marcela, 2010). These compounds provide onion with the characteristic odour and flavour, as well as most of their biological properties (Lanzotti, 2006). Organo-sulphur compounds present in onion preparations depend on the variety (Marcela, 2010) and the extraction and/or processing conditions.

Carbohydrates in onion constitute about 80% of the dry matter, and the major non-structural carbohydrates of the onion bulb are fructo-oligosaccharides, well known as fructan, followed by glucose, fructose and sucrose. The sugar concentration is associated with dormancy and decreases in glucose, fructose and fructan occur during storage, particularly after extended storage. Bacterial soft rot caused by *Erwinia* and *Pseudomonas* is the main postharvest disease when the bulbs are infected with the bacteria and stored at room temperature (Cho, 2010).

Various protein sources can be found in intact onion in low concentrations, including lectins (the most abundant proteins in onion), and other classes include prostaglandins, fructan, pectin, adenosine, vitamins (B1, B2, B6, C, E, biotin and nicotinic acid), fatty acids, glycolipids, phospholipids and essential amino acids.

Bioactive components have been studied for several decades (Rodrigues, 2009) and the importance of biological and pharmacological activities, such as antifungal, antibacterial, antitumor, anti-inflammatory, antithrombotic and hypocholesterolemic properties of certain steroid saponins and sapogenins, such as b-chlorogenin, has been demonstrated (Lanzotti, 2006). These chemical compounds are reported to exhibit several biological effects, including cholesterol reduction, cancer prevention and others, and probably work synergistically with organo-sulphur compounds (Corzo-Martínez, 2007)

Recent attention has focused on the polyphenols that are also responsible for a great part of the health benefits. Several thousand molecules having a polyphenol structure (ie, several hydroxyl groups on aromatic rings). The polyphenols have been identified in higher plants, and several hundred are found in edible plants, such as the onions. These molecules are secondary metabolites of plants and are generally involved in defense against ultraviolet radiation or aggression by pathogens. These compounds may be classified into different groups as a function of the number of phenol rings that they contain and of the structural elements that bind these rings to one another. Distinctions are thus made between the phenolic acids, flavonoids, stilbenes and lignans. The flavonoids, which share a common structure consisting of 2 aromatic rings (A and B) that are bound together by 3 carbon atoms that form an oxygenated heterocycle (ring C), may themselves be divided into 6 subclasses as a function of the type of heterocycle involved: flavonols, flavones, isoflavones, flavanones, anthocyanidins and flavanols (catechins and proanthocyanidins). In addition to this diversity, polyphenols may be associated with various carbohydrates and organic acids and with one another. Flavonols are the most ubiquitous flavonoids in foods, and the main representatives are quercetin and kaempferol. The richest sources of flavonols are the onions (up to 1.2 g/kg fresh wt). These compounds are present in glycosylated forms. The associated sugar moiety is very often glucose or rhamnose, but other sugars may also be involved (eg, galactose, arabinose, xylose, glucuronic acid). These flavonols accumulate in the outer and aerial tissues (skin and leaves) because their biosynthesis is stimulated by light. Marked differences in concentration exist between pieces of different onions, even in the same plant, depending on exposure to sunlight (Price, 1995).

For these reasons, onions are one of the major food source of dietary flavonoids, with average content ranging from 270 to 1917 mg of flavonoids per Kg of fresh weight, depending on the onion variety. For example, the red onions consumed by the Brazilian population have over 1000 mg/kg (Arabbi et al., 2004). They are mainly represented by the flavonols quercetin and kaempferol, as indicated, but it can also find other flavonoids such as isorhamnetin, myricetin and their conjugates (as glycosylated forms) (Slimestad, 2007). Of all these, quercetin and its conjugates are the predominant flavonols present. The abaxial epidermis of the scales contained a higher level of flavonols than did the mesophyll and approximately 50% of flavonols were detected in the top quarter part of the scales. Anthocyanins, namely peonidin 3-glucoside, cyanidin 3-glucoside and cyanidin 3-arabinoside and their malonylated derivatives, cyanidin 3-laminariobioside and delphinidin and petunidin derivatives, are located in the skin of red onions and in the outer fleshy layer (Donner, 1997, Škerget, 2009). It is known that quercetin concentration was higher in red onions than in the white variety (Gorinstein, 2008). Although this bulb grows partially underground in the skin dominates by two types of quercetin, quercetin aglycone and quercetin-4-glocusid (Suh et al., 1999, Ly et al., 2005). Quercetin is modified by an oxidation process which provides the characteristic brown color. These products are 3,4-dihydroxybenzoic acid, and 2,4,6-trihydroxyphenylglycosilic acid, which also offer antimicrobial protection (Takahma and Hirota, 2000, Takahama et al., 2001).

The extraction of flavonoids depends on their solubility and is governed by their chemical nature. There is a possibility of interaction of flavonoids with other plant components such as carbohydrates and proteins. These interactions may lead to the formation of complexes that may be quite insoluble. The solubility of flavonoids is also affected by the polarity of the solvents used for extraction. Solvents, such as methanol, ethanol, propanol, acetone, ethyl acetate, dimethylformamide and their combinations have been used for the extraction of flavonoids, often in combination with different proportions of water. The recovery of polyphenols from plant materials is also influenced by the extraction time and related factors. Reported extraction periods vary from 1 min to 24 h (Naczk, 2006). Longer extraction times increase the chance of oxidation of phenolics unless reducing agents are added to the solvent system.

The extraction also can be influenced by the ratio of solvent-to-sample. The recovery from dry onion was strongly influenced by variations in the sample particle size.

Application of subcritical water extraction (SWE) for isolation of bioactive phytochemicals from onion is an alternative to the use of solvents. The SWE is carried out using hot water, above its boiling point, that is sufficiently pressurized to 15 bar at 200 ºC and 85 bar at 300 ºC, to maintain a liquid state. The efficiency of SWE is affected by temperature and extraction time as well as by the presence of small quantities of organic solvents and surfactants. The polarity of water under pressure varies with temperature. At lower temperatures the water is a very polar liquid but at higher temperatures (≥250 °C) the polarity of pressurized water is similar to that displayed by polar organic solvents. Thus, the solubility of analytes in superheated water can be manipulated by changing the extraction temperature. Under subcritical conditions the polar analytes are selectively extracted at lower temperatures while less polar analytes are extracted at higher temperatures. SWE is a new technique that can be applied for the extraction of antioxidants from onion (Naczk, 2006).

Absorption and Metabolism

There are several factors affecting flavonoid absorption, like:

- The presence or absence of glycosylation on hydroxyl groups: The glycosylated form of flavonols are absorbed better than the aglycone form. Studies in people have found that most of the glycosidic quercetin has been absorbed (Walle, 2000). It is a fact that flavonoid aglycones are hydrophobic and as they are glycosylated this hydrophobicity vanishes (Hollman, 1995). The presence of the glycosidic group, suggests that the glycosidic quercetin is hydrolyzed by β-glucosidases and are better absorbed (Walle, 2000).
- The position of glycosylation: The position of the sugar attached to the flavonoid affects absorption in the intestine. Between quercetin-3-glucoside and quercetin-4'-glucoside there is not difference in absorption in humans (Olthof, 2000). In contrast between quercetin-4'-glucoside and quercetin-3-rutinoside it is observed that there is a greater absorption of the first with respect to the second. Moreover, the time of appearance in plasma is lower for quercetin-4'-glucoside, suggesting that this is absorbed in the small intestine and the other in the colon (Hollman and Katan, 1999)
- The quality of sugar moiety attached: For quercetin it has been shown in rats that the glycosylated form of it is better absorbed than the aglycone form and is also influenced by the type of sugar. The tests were performed with quercetin aglycone, quercetin-3-rhamnosida or rutin and quercetin-3-glucoside, and the best absorption was that of quercetin-3-glucoside (Morand, 2000). This shows the difference in

the absorption time can vary from 0.5h to 9 h, depending on the type and quality of sugar attached to the flavonoid (Aziz, 1998)

- Plant/food matrix: Studies have shown that quercetin contained in onions is better absorbed than that contained in apples or supplements. In apples there are a variety of groups attached to the quercetin glycosides which interfere with the rate of absorption compared with the onion (Goldberg, 2003).
- Interactions with proteins, micelles, and emulsifiers: The absorption of quercetin is strongly related to the solubility (Piskula and Terao, 1998).

The absorption is also affected by the substances accompanying the flavonoid. Quercetin is highly soluble in combination lipids and emulsifiers . Various combinations have been tested and the best is the soybean oil and sucrose fatty acid ester. In addition, the lipid micelles dispersion can be an important factor for the best absorption (Azuma, 2002). With regard to the presence of protein, they can affect (Carbonaro et al., 2001) or not (Hollman et al., 2001) the absorption of flavonoids. The mixture of milk with tea seems to have no effect with antioxidant activity (Leenen et al., 2000).

On the other hand, the absorption of flavonoids could be occur in two ways: via passive or active way. Passive absorption of flavonoid aglycone can occur across a membrane by diffusion. As for the glycosidic flavonoids, they can be transported by actively, since they are hydrophilic and may use the form of glucose uptake although this has not been tested (Hollman, 1995).

The metabolism of flavonoids differ at different stages from the mouth to the colon. In the mouth, enzymes catalyze the hydrolysis of glycosidic flavonoid. The ingestion of glycosylated quercetin; quercetin may be observed free in saliva, as well as the inoculation of saliva samples of quercetin-4'-glucoside causes the release of quercetin in minutes but this does not happen with rutin (Hirota, 2001) .

The continuous absorption in the small intestine from the lumen to the epithelial cells which are attached to a series of enzymes that hydrolyze the glycosylated flavonoid (Day, 2000; Depient, 2002) or you can go through the enterocyte via the active way following an intracellular hdrolysis (Hollman, 1995). Flavonoids free in the lumen can enter the enterocyte by passive means. Using the human intestinal cell model Caco-2 , shows that there is good absorption of quercetin aglycone but not their glucosides (Muroto, 2000). The process of absorption also depends on the quality of the flavonoid aglycone and the position of the sugar attached to it. Absorption of quercetin-4'-

glucoside depends on two interactions: the sodium-dependent glucose transporter and hydrolysis by lactase phlorizin hydrolase-(LPH), but quercetin-3-glucosid only needs the LPH (Day, 2003).

On the other hand, it has been shown that the quercetin-4'-glucoside, naringenin-7-glucoside, apigenin-7-glucoside, genistein-7-glucoside, and daidzein-7-glucoside were rapidly deglycosylated in the small intestine and human liver cells , while quercetin-3, 4-glucoside, quercetin-3-glucoside, kaempferol-3-glucoside, quercetin-3-rhamnoglucoside and naringenin-7-rhamnoglucoside were not hydrolyzed (Day, 1998). Finally, all the flavonoids that were not absorbed before, reaching the colon. In it there are plenty of microflora which act on these and are hydrolyzed for subsequent absorption (Manach, 1998; Hollman, 1995, Mc Donald, 1983). In the colon, many reactions occur. For example, the rutin, not hydrolysed in the small intestine, is deglycosylated by the bacterial microflora. In addition, there are fissures in the rings of flavonoids and phenolic acids formed. Most of the flavonoid aglycones become glucose and then absorbed by the body. This enzymatic process is produced by bacteria (Kim, 1998).

After having performed its functions in the body, flavonoids are excreted using two ways: urine and bile. When they are excreted in bile, they reach the small intestine and then go to the colon where they are metabolized and the metabolites are again reabsorbed by the body. Flavonoids that are used in the liver are eliminated in the urine (Hackett, 1986; Griffiths, 1982)

ALTERATION OF FLAVONOIDS

Now it will be shown what are the effects of the different variables that can affect the flavonoid content in foods:

Storage: Storage may significantly affect the amount of flavonoids in food, as these may undergo oxidation. Sometimes this oxidation is necessary for example in tea, but this oxidation causes to lose much of the flavonoids by the formation of polymers (Susulski, 1982). However, the cold storage does not seem to affect the quality and content of flavonoids in onions (Price, 1997). Moreover, this degradation effect depends on the type of flavonoid content in the food. At -20 ° C for 9 months quercetin varied by only a small amount in onions , while the levels of quercetin decreased by 40% in bilberries and increased 32% in strawberries, during the same time (deVries, 1998; Price, 1997).

Preparation and cooking: the shelling of a food, fruit or vegetable, can affect the flavonoid content, because many of these are found in the skin where they are synthesized (Manach, 2004). In the case of onions, peeling can lose 21% of quercetin-4'-glucoside and 73% of antiocianinas (Gennaro, 2002) Cooking can reduce the amount of quercetin in onions by 75 % to 80% if they are boiled for 15 min. In the microwave it can be reduced by 65% and 30% during frying (Crozier, 1997). Usually boiling is the great destroyer of the flavonoids, mainly quercetin, present in plants, where the cooking water contains this compound (Crozier, 1997; Ewald, 1999; Price, 1997; Manach, 1998), which is why onion soup has become a good source of flavonoids (Nemeth, 2004; Takenaka, 2004). It is a fact that cooking removes much of the antioxidant activity and this decrease was observed in the antihypertensive activity in rats (Agostini, 2004; Kawamoto, 2004). Moreover, the addition of seasonings and the presence of iron facilitates and accelerates the hydrolysis of quercetin glycosides (Ioku et al., 2001).

Processing: Non-enzymatic browning reactions can occur during processing causing deterioration in the quality of the onion through the color changes (Kaymak-Ertekin and Gedik, 2005, Kumar et al., 2005). The colours pink or blue-green can occur after processing (Kubec et al., 2004, Imai et al., 2006). Moreover, the processing can increase the amount of the flavonoid quercetin in onions. The process of pasteurization and fermentation can free quercetin aglycone or carry out enzymatic hydrolysis which results in an increase in the amount present of this flavonoid (Stewart et al., 2000, Price and Rhodes, 1997).

Antioxidant Properties and Health Benefits

There are two classes of antioxidants, endogenous and exogenous. More of the last ones are from the plant cells because they contain a complex array of components such as vitamin C, vitamin E, glutathione and phenolics (including flavonoids) which offer protection against cellular damage. (Hodges et al., 1999)

Ingestion of such compounds may also offer direct chemoprotective roles in animal cells and help reduce oxidative stress. In addition, many of these compounds may also initiate the animal cells to produce their own chemical oxidative defence mechanisms (Teyssier et al., 2001)

Several components of the onion have antioxidant effects. The sulfur compounds, such as S-allylcysteine, diallyl sulfide, diallyl disulfide, Sethylcysteine, and N-acetylcysteine can inhibit lipid oxidation (Imai et al., 1994, Ou et al., 2003, Huang et al., 2004). Allicin also reduces the amount of hydroxyl radical, acting as a cleanser (Prasad et al., 1995). These effects correspond with the increase of the enzymes catalase, superoxide dismutase and glutathione peroxidase (Xiao and Parkin, 2002).

It is known that the presence of quercetin and kaempferol in onions are inhibitors of lipid oxidation also their aglycone form is better than the glycosylated form; the reason is the presence of free hydroxyl groups. The antioxidant activity may be for cleaning or metal chelation (Santas et al., 2008).

Studies *in vitro* have associated the content of flavonoids and phenolic compounds (in several species of onions) with the antioxidant effect, finding differences between them. Also the color of the species is related to the amount of antioxidants found, greater amounts in red and yellow onions (Yang et al. 2004; Stratil et al., 2006). The color in these plants is related to an antioxidant compound anthocyanin (Griffiths et al., 2002). Moreover, some recent studies have evaluated the potential effectiveness of onion and its bioactive compounds as natural preservatives. It has been described that the addition of onion lyophilized powder can be effective in delaying the lipid oxidation in oil-in-water emulsions which can be considered a good model of food emulsion, such as mayonnaise and margarine (Santas et al., 2008).

The antioxidant compounds in onion are as effective as "vitamin E" when used to counteract the effects of cancer induced in rats with nicotine (Helen et al., 2000). Otherwise, 5.5 servings of onions, is equivalent to 2 cups of tea, 4 apples, 3.5 glasses of blackcurrant juice, 7 cups of orange or 20 glasses of apple juice, if it is measured by TEAC assay (Paganga et al., 1999).

Santas et al. studied the antioxidant effect of different fractions of three different Spanish onion varieties.. The ethyl acetate fraction and water subfractions of methanolic extract were analyzed. Flavonoids were mainly present in ethyl acetate subfraction being 34.92 ± 0.75, 7.95 ± 0.16, 0.38 ± 0.01 micromol of rutin eq./g D.W. and its antioxidant capacity was 74.86 ± 1.77, 24.59 ± 0.67, 4.55 ± 0.44 micromol Trolox / g D.W. of Grano de Oro, Fuentes de Ebro and Calçot de Valls varieties, respectively (Santas, 2010).

Antifungal, Antiparasitic and Antibacterial Properties

It is known that the plant genus *Allium* is associated with the treatment of infections (Farbman et al., 1993), which is why we have focused research into these therapeutic effects and to try to describe them (Harris et al., 2001, Sivam, 2001).

The antibacterial properties of the onion are related to bacteria that cause tooth decay, Streptococcus mutans and S. sobrinus, in addition to causing adult periodontitis, Porphyromonas gingivalis and Prevotella intermedia (Kim, 1997). Studies have shown that these antibacterial properties are related to sulfur-containing components in onions (Peter et al., 2005). Bacteria such as Bacillus, Micrococcus, Staphylococcus, Streptococcus, Salmonella and Escherichia coli enteridis are sensitive to these compounds (Corzo et al. 2007; Majheniě et al., 2009). It has also been found that components of Allium sulfur are effective on the bacterium Helicobacter pylori, which is related to the development of stomach cancer, consumption of allium vegetables may reduce the risk of gastric cancer (You et al. 1998).

Studies carried out with quercetin and kaempferol extracted onion have shown that quercetin is more effective than kaempferol in the disc diffusion assay. The kaempferol was effective in inhibiting the growth of Staphylococcus aureus, Micrococcus luteus, Bacillus cereus, Listeria monocytogenes and Pseudomonas aeruginosa. Quercetin, however showed activity in all the above plus bacteria Escherichia coli, in addition, it was also found that gram-negative bacteria are more resistant than gram-positive (Santas et al., 2010). The modus operandi of these flavonoids are inhibition of DNA, RNA and protein biosynthesis, which causes cytoplasmic membrane functions to be altered (Cushnie and Lamb, 2005). It has also been observed that the aglycone form are more effective than the glycosylated forms, due to the absence of C-3 hydroxyl group. It means that the aqueous extracts of these flavonoids are less effective because the glycosylated forms are more soluble in water (Xu and Lee, 2001).

Different studies claim that the crude extracts of onions have antifungal properties (Kim, 1997; Yin and Tsao, 1999). Onion oil is effective against dermatophyte fungi (Zohri et al., 1995), fungi sensitive to these components include: Candida, Trichophyton, Torulopsis, Rhodotorula, Cryptococcus, Aspergillus, and Trichosporon (Peter et al., 2005). This is because the extracts have the capacity to inhibit the synthesis of lipids, proteins and nucleic acids

(Adetumbi et al., 1986). On the other hand, it has also been shown the inhibition of the growth and metabolism of certain parasites (Mon et al., 1994). Various types of Leishmania are susceptible to extracts of onions and it is suggested by the sulfur components contained in this (Saleh et al., 2004).

Clearly there are many claims on the health benefits of onions. However, most have not received any rigorous scientific investigation. Here we report some of the advances in defining the health benefits of onion where the mechanisms behind these effects are known.

ANTIARCINOGENIC PROPERTIES

The content of sulfur compounds in onions has proved effective in protecting against cancer. In studies in humans they have been found to reduce the risk of stomach, esophageal and prostate cancer. This research was carried out in China's population who consumed vegetables from the allium genus such as onions (Gao et al., 1999, Hsing et al., 2002). It also reduces the risk of developing cancer of the brain (Hu et al., 1999). A study reinforcing these findings was made in the Netherlands, which concluded that consumption of onions reduces the risk of developing stomach cancer (Dorantes et al., 1996). In France, a study has revealed that the risk of breast cancer in women decreases with the consumption of onions, garlic and fibers (Challier et al., 1998). In contrast, the study in Holland finds no link between the consumption of onions and garlic supplements with reduced risk of breast cancer. Moreover, there have been several studies in animals (rats) to verify and deepen understanding, of the relationship between the protective ability of sulfur compounds in onions and reduction of tumors induced in the stomach. The mechanisms of action are obtained (Sumiyoshi and Wargovich, 1990).

EFFECTS ON LIPID METABOLISM

The work carried out with the onion has shown that (+)-S-methyl-L-cysteine sulphoside possesses a hypolipidaemic effect. This compound reduces the amount of cholesterol, triglycerides and phospholipids, and in addition this decrease is higher with the use of some medications (Kumari et al., 1995). Quercetin also has features to inhibit hepatic cholesterol biosynthesis in vitro (Glasser et al., 2002). Consuming onion has succeeded in reducing the index of biliary cholesterol and its index of hydrophobicity. In addition the values in serum and liver were decreased in studies with mice after inducing a state of severity of biliary cholesterol. Hepatic hydroxymethylglutaryl coenzyme A reductase activity was increased after consumption of onions, then observed a degradation of cholesterol by an enzyme. These results confirm that degradation of cholesterol is accelerated when mice are fed with onions (Vidyashankar et al., 2010). Furthermore, *in vitro* studies with onions indicate that possess antioxidant activity on LDL oxidation. (Higuchi et al., 2003).

Also, it is known that oil extracts derived from *Allium cepa* onions are as effective as the use of vitamin E, to increase resistance in the nicotine-induced lipid peroxidation in rats (Helen et al., 2000).

Cardiovascular Effects

The derivative of allicin thiosulfinates allium is an inhibitor of hypertension by reducing serum cholesterol and triglyceride levels. In the form of a drug called Captopril it has been successfully used to prevent hypertension in rats. These two substances have significantly decreased triglycerides and blood pressure (Miron et al., 2004). The compounds of the type α-sulphilnyl-di-sulphides (cepaenes) have anti thrombotic activity, therefor, it is considered to be the sulfur in the onion compounds that are responsible for antiplatelet activity (Block et al., 1997).

It is known that platelet aggregation is one of the causes of heart attacks and stroke, which is why it was tried to use compounds for treating these ailments (Orvis et al., 2001). Often we find studies with contradictory results. Ali et al., (1999) were unable to demonstrate an antiaggregatory effect of onion on human blood platelets while in an earlier study (Srivastava, 1984) a very strong effect was reported.

Later it was explained that aqueous extracts of onion suggest the release of arachidonic acid, because it starts the metabolism of eicosanoids, which inhibit the synthesis of thromboxane (Moon et al., 2000). Studies have been carried out with fresh onion juice, which when placed in human blood has induced platelet aggregation. The results show that the juices inhibit platelet aggregation and that this inhibition becomes smaller with time (Osmont et al., 2003). The effectiveness of platelet aggregation with sulfur compounds in onion is affected by the type of cooking. The maximum benefit occurs when eaten raw and heavy drinking can also cause a pro-aggregate effect. (Cavagnaro et al., 2007).

Extracts of Welsh onion (*Allium fistulosum L*) can also modulate vascular responses in aortae (Chen et al., 1999). Raw extracts of this species lower systolic blood pressure and prolong bleed times in rats possibly through inhibition of platelet function and the suppression of thromboxane production (Chen et al., 2000).

Quercetin also seems to have antiplatelet effects. In a pilot study in humans was provided quercetin-4'-glucoside,. Then measured values of this testing for the presence in plasma were done.

Subsequently, tests were applied to determine platelet aggregation in the samples and the results showed that there is an antiplatelet effect after 30 min (Hubbard et al., 2004). Supporting this research, a study in humans found that onion soup inhibits some aspects of platelet aggregation collagen-stimulated ex vivo (Hubbard et al., 2006).

Bone Effects

The effect of fruit and vegetables on bone are being studied. Experiments suggest that eating certain fruits and vegetables improves the absorption of calcium contained in milk. The sulforaphane, an inducer of enzymes found in plant cells, has an effect on joints.

The flavonoids in onions as active phytochemicals act to bone health. Moreover, you must eat fruits and vegetables combined so that there is synergy between them in order to enhance their effect on bone (Lister et al., 2007).

In a study in postmenopausal non-Hispanic women, to determine the relationship between onion consumption and bone density, it has been found that eating onions once a day, or more, gives a 5% increase in bone density. Therefore, the researchers concluded that consumption of onions have effects on bone density in perimenopausal and postmenopausal women with non-Hispanic 50 years, besides the risk of bone fracture was reduced by 20% compared to those who never consumed onions (Matheson et al., 2009)

Anti-Inflammatory Properties

Various diseases such as colon cancer, atherosclerosis, inflammatory bowel disease, multiple sclerosis, Alzheimer's disease and septic shock are related to the production of prostaglandins and nitric oxide. These compounds exert a regulatory role in inflammatory processes of the human body. When there is an alteration in the production of these compounds, they can trigger the diseases described above (Moncada and Higgs,1991; Claria, 2003; Ali et al., 2000). Recent studies have shown that antioxidants can inhibit the formation of nuclear factor kappa B, which is regulated by prostaglandins and nitrous oxide, which means they can reduce the symptoms from these diseases. So you can find the beneficial relationship between the consumption of onions, cardiovascular disease and cancer. It has also been found that extracts of onion inhibit cyclooxygenase and lipoxygenase, which are related to the diseases described above (Srivastava, 1986).

Some of these properties are attributed to diallyl disulfide, a sulfur component found in onions, which can significantly reduce the secretion of proinflammatory, such as cytokines (Keiss, 2003).

OTHER EFFECTS

Effects on the Respiratory System

Cepaenes and Thiosulphinates, compounds derived from onions, have antiasthmatic activity (Dorsch et al. 1989; Dorsch et al., 1996), which inhibit the effects of cyclooxygenase and lipoxygenase in eicosanoid metabolism and cause bronchial restriction (Wagner et al., 1990).

Selenium Complement

Nowadays onion is found in a wide range of processed foods particularly in the so-called fast foods. In this context, it is a good candidate for a delivery system for beneficial mineral nutrients such as selenium, making it a potential 'functional food'.

This could have major health benefits for the population in the areas of mental health, fertility (both male and female), cancer and cardiovascular disorders (reviewed Rayman, 2000).

It has been demonstrated that growing onions in a selenium-rich environment can elevate the endogenous amount to levels that can deliver 1–3 RDA equivalents in a standard 80 g serving (Kopsell and Randle, 1997).

Antidepressant

At present there is still some considerable debate as to what component(s) of onion are essential for delivering health benefits. Some attempts to purify extracts of the antidepressant hypericin and administer it in pure form have met with less success than the cruder extract (Linde et al., 1996). This may be due to a synergistic effect between many of the components in the extract that result in an overall health benefit.

Inulin (a type of fructan) rich in chicory, artichoke, garlic, leek and onions also improved blood lipid profiles and altered the colonic environment in a beneficial manner and shows promise as a functional ingredient in many processed foods (Causey et al., 2000).

OTHER USES: FOOD PRESERVATION

Onions and onion by-products have been proposed as food ingredients which can be added to food preparations to enhance their shelf-life preservation.

It has been described that peeling of onions significantly decreases their total flavonoid content and antioxidant capacity (approximately half the starting level) due to the fact that an important amount of flavonoids is concentrated in the outer parts of the onion (Lee and Lee et al., 2008; Santas, 2010).

However, there are some concerns regarding its use in food conservation. On the one hand, the stability of onion antimicrobial and antioxidant compounds during storage until its further use as a food ingredient is an important issue. Minimal losses of onion antioxidant capacities have been reported in frozen and canned onions during long periods of time (up to 8 months) especially under the presence of ascorbic acid (Murcia and Jiménez et al., 2009).

On the other part, it is questionable whether onion bioactive compounds would remain stable after their processing and addition as food ingredients to processed or cooked foods or preparations. Great attention has been focused on processing at high temperatures as it is an important factor which can induce significant changes in the concentration, antioxidant properties and bioavailability of onion bioactive compounds (Rohn and Buchner et al., 2007). Nevertheless, further processing by blanching, moderate microwave heating, frying with either rape-seed oil or butter has a lower influence on total phenolic or flavonoid content (Gorinstein and Leontowicz et al., 2008; Pellegrini and Miglio et al., 2009). In contrast, it has been described that after treatments at high temperatures such as roasting (180ºC), quercetin glycosides are strongly degraded leading to their deglycosilation and the formation of the quercetin aglycon (Rohn and Buchner et al., 2007; Rodrigues and Perez-Gregorio et al., 2009). The quercetin aglycon is heat stable (Rohn and Buchner et al., 2007).

In the case of boiling water (100ºC) there is a significant decrease in onion flavonol content and its capacity to inhibit lipid peroxidation due to the solubilisation of flavonols in water (Makris and Rossiter, 2001; Murcia and Jiménez-Monreal, 2001).

Special attention has been focused on the utility of onion as a natural ingredient to delay the deterioration of meat products as it is well known that the preservation of its palatability, taste, colour and flavour without adding synthetic preservatives is highly demanded by customers (Santas, 2010). Some authors have reported that the addition of onion powder to pork or beef refrigerated meat, delayed the formation of primary and secondary oxidation products (Park and Yoo et al., 2008) and inhibited the growth of food spoiling microorganisms, such as Enterobacteriaceae (Irkin and Arslan, 2010; Park and Yoo et al., 2008). In addition, the effect of juice extracts from onions on lipid oxidation and sensory quality in refrigerated turkey meat has been analyzed (Tang and Cronin, 2007). They not only improved the taste, but also provided protection against oxidation during cooking and refrigerated storage. Furthermore, onions have been described as being a good additive to decrease the formation of nitrogen heterocycles, such as polycyclic aromatic hydrocarbons and azaarenes, which can be formed during cooking practices and can act as human carcinogens (Janoszka, 2010). Finally, onion bioactive compounds can also be useful antibrowning agents preventing other vegetables and fruits from the deterioration associated with the post-harvest process (Roldan and Sanchez-Moreno et al., 2008). On balance, onion can be considered a good food preservative, especially in fresh or minimally processed foods, and can play a crucial role when conveniently combined with other food preservation technologies (i.e. dehydratation or heat treatment).

Conclusion

In summary, onions are a good and useful ingredient for human health and disease prevention, but there is the danger that this may lead to people substituting good and varied eating practices with poor nutrition, very rich in onion or even with supplements in the form of pills.

A balanced diet with onions and other vegetables in an unprocessed or cooked form, along with a healthy lifestyle are a combination that will help prevent many of the Western diseases.

References

Agostini, L.R., Morón, M.J., Ramón, A.N., Ayala, A. (2004). Determination of the antioxidant capacity of flavonoids in fruits and fresh and thermicaly treated vegetables. Arch. Latinoam. *Nutr.* 54,89–92.

Ali, M., Thomson, M., Afzal, M. (2000). Garlic and onions: their effect on eicosanoid metabolism and its clinical relevance. *Prostaglandins LeukotEssent Fatty Acids,* 62, 55-73.

Antonia Murcia, M., Jiménez, A.M., Martínez-Tomé, M. (2009). Vegetables antioxidant losses during industrial processing and refrigerated storage. *Food Res. Int.*, 42, 1046-1052.

Arabbi, P.R., Genovese, M.I., Lajolo, F.M. (2004). Flavonoids in vegetable foods commonly consumed in Brazil and estimated ingestionby the Brazilian population.*J. Agric. Food Chem.* 52, 1124-31.

Azuma, K., Ippoushi, K., Ito, H., Higashio, H., Terao, J. (2002). Combination of lipids and emulsifiers enhances the absorption of orally administered quercetin in rats. *J. Agric. Food Chem.* 50, 1706–1712.

Block, E., Gulati, H., Putman, D., Sha, D., Niannian, Y., Zhao,S-H. (1997). Allium Chemistry: synthesis of 1-[alk(en)ylsulfinyl]-propyl alken(en)yl disulfides (cepaenes), antithrombotic flavorants from homogenates of onion (*Allium cepa*). *J. Agric. Food Chem.*45, 4414-4422.

Burda, S., Oleszek, W., Lee, C.Y. (1990). Phenolic compounds and their changes in apples during maturation and cold storage. J. Agric. Food Chem., 38,945–948.

Carbonaro, M., Grant, G., Pusztai, A. (2001).Evaluation of polyphenol bioavailability in rat small intestine. *Eur. J. Nutr.* 40, 84–90.

Cavagnaro, P.F., Camargo, A., Galmarini, C.R., Simon, P.W. (2007)Effect of cooking on garlic (*Allium sativum L.)* antiplatelet activity and thiosulfinates content. *J. Agric. Food Chem.* 55, 1280–1288.

Cavagnaro, P.F.,Sance, M.M.,Galmarini, C.R. (2007).Effect of heating on onion (*Allium cepa L.*) antiplatelet activity and pungency sensory perception. *Food science and technology international*, 13, 447-453.

Cavallito C.J., Bailey J.H. (1944). Allicin, the antibacterial principle of *Allium Sativum* i. isolation, physical properties and antibacterial action. *J. Am. Chem. Soc.*, 66, 1950–1951.

Challier, B., Perarnau, J.M., Viel, F.J. (1998). Garlic, onion and cereal fibre as protective factors for breast cáncer: a French case-control study. *Eur. J. Epidemiol,* 14, 737-747.

Cho, J.,Bae, R.N., Lee, S.K. (2010).Current Research Status of Postharvest Technology of Onion (*Allium cepa L.*). *Korean Journal of Horticultural Science and Technology,* 28, 522-527.

Clària, J. (2003).Cyclooxygenase-2 biology.*Curr. Pharm.* Des.,9, 2177-2190.

Corzo-Martínez, M., Corzo, N.,Villamiel, M. (2007).Biological properties of onions and garlic. *Trends Food Sci. Technol,* 18, 609-625.

Crozier, A., Lean, M.E.J., McDonald, M.S., Black, C.(1997). Quantitative analysis of the flavonoid content of commercial tomatoes, onions, lettuce, and celery. *J. Agric. Food Chem,* 45, 590–595.

Cushnie, T.P.T., Lamb, A.J. (2005).Antimicrobial activity of flavonoids. Int. *J. Antimicrob. Agents,* 26, 343-356.

Day, A., Geeb, J.M.,DuPontb, M.S., Johnsonb, I.T., Williamsonb, G. (2003). Absorption of quercetin-3-glucoside and quercetin-4'-glucoside in the rat small intestine: the role of lactase phlorizin hydrolase and the sodium-dependent glucose transporter. Biochem.*Pharmacol.* 65, 1199–1206.

Day, A.J., Canada, F.J., Diaz, J.C. (2000). Dietary flavonoid and isoflavone glycosides are hydrolysed by the lactase site of lactase phlorizin hydrolase. *FEBS Letters*, 468, 166-170.

Day, A.J., DuPont, M.S., Ridley, S. (1998).Deglycosylation of flavonoid and isoflavonoid glycosides by human small intestine and liver beta-glucosidase activity. *FEBS Letters*,436, 71-75.

Degirmencioglu, N., Irkin, R. (2009).Survival of some microorganisms in the presence of onion (*Allium cepa L.*) extracts in vitro. *Pol. J. Food Nutr. Sci., 59,* 61-66.

Depeint, F., Gee, J.M., Williamson, G., Johnson, I.T. (2002).Evidence for consistent patterns between flavonoid structures and cellular activities. *Proc. Nutr. Soc.* 61, 97–103.

deVries,J., Hollman, P.C.H., Meyboom, S.(1998). Plasma concentrations and urinary excretion of the antioxidant flavonolsquercetin and kaempferol as biomarkers for dietary intake. *Am. J. ClinNutr,* 68, 60.

Donnerb, H., Gaob, L., Mazzab, G. (1997).Separation and characterization of simple and malonylatedanthocyanins in red onions, *Allium cepa L. Food Res. Int.* 30,637-643.

Dorant, E., van den Brandt, P.A., Goldbohm, R.A., Sturmans, F. (1996).Consumption of onion and a reduced risk of stomach carcinoma. *Gastroenterology,* 110, 12-20.

Ewald, C., Fjelkner-Modig, S., Johansson, K., Sjoholm, I., Akesson, B.(1999). Effect of processing on major flavonoids in processed onions, green beans, and peas. *Food Chem*, 64, 231.

Farbman, K.S., Barnett, E.D., Bolduc, G.R., Klein, J.O. (1993). Antibacterial activity of garlic and onions: a historical perspective. Pediatr. *Infect. Dis. J.,* 12, 613-614.

FromtNug, R. A., Bulmer,G. S. (1978). In vitro effect of aqueous extract of garlic (*Allium sativum*) on the growth and viability of Cryptococcus neoformans.*Mycologia*, 70, 397-405.

Gao, C.M., Takezaki, T., Ding, J.H., Li, M.S., Tajima, K. (1999). Protective effect of allium vegetables against both esophageal and stomach cancer: a simultaneous case-referent study of a high-epidemic area in Jiangsu Province, China. *Jpn. J. Cancer Res*. 90, 614–621.

Gennaro, L., Leonardi, C., Esposito, F., Salucci, M., Maiani, G., Quaglia, G., Fogliano, V. (2002). Flavonoid and carbohydrate contents in Tropea red onions: effects of homelike peeling and storage. *J. Agric. Food Chem.* 50,1904–1910.

Glasser, G., Graefe, E.U., Struck, F., Veit, M., Gebhardt, R. (2002).Comparison of antioxidative capacities and inhibitory effects on cholesterol biosynthesis of quercetin and potential metabolites. *Phytomedicine*, 9, 33-40.

Goldberg, D.M., Yan, J., Soleas, G.J. (2003). Absorption of three wine-related polyphenols in three different matrices by healthy subjects.*Clin.Biochem.* 36,79–87.

Gorinstein, S., Leontowicz, H., Leontowicz, M., Namiesnik, J., Najman, K., Drzewiecki, J., Cvikrová, M., Martincová, O., Katrich, E., Trakhtenberg, S. (2008).Comparison of the main bioactive compounds and antioxidant activities in garlic and white and red onions after treatment protocols. *J. Agric. Food Chem.* 56, 4418-4426.

Griffiths, G., Trueman, L., Crowther, T., Brian Thomas, B., Smith, B. (2002).Onions—A Global Benefit to Health.*Phytother. Res.,* 16, 603–615.

Griffiths, L.A. (1982). Mammalian metabolismof flavonoids. In J.B. Harborne and T.J. Mabry (Eds.), The Flavonoids: Advances in Research (pp. 681-718). London: Chapman and Hall.

Hackett, A.(1986). The metabolism of flavonoid compounds in mammals. In V. Cody, E. Middleton, J.B. Harborne (Eds.), Plant flavonoids in biology and medicine: biochemical, pharmacological, and structure–activity relationships (pp. 177). New York: Alan R. Liss.

Harris, J.C., Cottrell, S., Plummer, S., Lloyd, D. (2001).Antimicrobial properties of *Allium sativum* (garlic). *Applied microbiology and biotechnology*, 57, 282-286.

Helen, A., Krishnakumar, K.,Vijayammal, P.L., Augusti, K.T. (2000).Antioxidant effect of onion oil (*Allium cepa L.*) on the damages induced by nicotine in rats as compared to alpha-tocopherol. *Toxicol. Letters,* 116, 61-68.

Higuchi, O.,Tateshita, K., Nishimura, H. (2003).Antioxidative activity of sulfur-containing compounds in allium species for human low-density lipoprotein (LDL) oxidation in vitro. *J. Agric. food chem.*, 51, 7208-7214.

Hirota, S., Nishioka, T., Shimoda, T., Miura, K., Ansai, T., Takahama, U. (2001). Quercetinglucosides are hydrolysed to quercetin in human oral cavity to participate in peroxidase-dependent scavenging of hydrogen peroxide. *Food Sci. Technol.* Res. 7, 239–245.

Hitokoto, H., Satoshi, L., Tomoaki, L., Sakai, S., Kurata, H. (1980).Inhibitory effects of spices on growth and toxin production of toxigenic fungi. *Applied and environmental microbiology*, 39, 818-822.

Hollman, P.C., Katan, M.B. (1999). The sugar moiety is a major determinant of the absorption of dietary flavonoid glycosides in man. *Free Radic. Res.,* 31, 569–573.

Hollman, P.C., Van Het Hof, K., Tijburg, L., Katan, M. (2001). Addition of milk does not affect the absorption of flavonols from tea in man. *Free Radic. Res.,* 34, 297–300.

Hollman, P.C.H., deVries, J.H., vanLeewen, S.D., Mengelers, M.J.B., Katan, M.B.(1995).Absorption of dietary quercetin glycosides and quercetin in healthy ileostomy volunteers. *Am. J. ClinNutr,* 62,1276.

Hsing,A. W., Chokkalingam,A. P., Gao,Y. T., Madigan,M. P., Deng,J., Gridley.G.,Fraumeni,J. F., Jr. (2002). Allium Vegetables and Risk of Prostate Cancer: A Population-Based Study. *Natl. Cancer Inst.* 94, 1648–1651.

Hu, J., La Vecchia, C., Negri, E., Chatenoud, L.,Bosetti, C., Jia, X., Liu,R., Huang, G., Bi, D., Wang, C. (1999). Diet and brain cancer in adults: A case-control study in Northeast China *Int. J. Cancer,* 81,20–23.

Huang, C.N., Horng, J.S., Yin, M.C. (2004).Antioxidative and antiglycative effects of six organosulfur compounds in low-density lipoprotein and plasma. *J. Agric. Food Chem.* 52, 3674-3678.

Hubbard, G.P., Wolffram, S., Lovegrove, J.A., Gibbins, J.M. (2004). Ingestion of quercetin inhibits platelet aggregation and essential components of the collagen-stimulated platelet activation pathway in humans. *J. ThrombHaemost*, 2, 2138–2145.

Hubbard, G.P., Wolffram, S., Lovegrove, J.A., Gibbins, J.M. (2004). Ingestion of quercetin inhibits platelet aggregation and essential components of the collagen-stimulated platelet activation pathway in humans. *J. ThrombHaemost*, 2, 2138–2145.

Imai, J., Ide, N., Nagae, S., Moriguchi, T., Matsuura, H., Itakura, Y. (1994). Antioxidant and radical scavenging effects of aged garlic extract and its constituents. *Planta Med.*,60, 417-420.

Imai, S., Akita, K., Tomotake, M., Sawada, H. (2006).Identification of two novel pigment precursors and a reddish-purple pigmentinvolved in the blue-green discoloration of onion and garlic.*J. Agric. Food Chem.* 54, 843-847.

Ioku, K.,Aoyama, Y.,Tokuno, A.,Terao, J.,Nakatani, N.,Takei, Y. (2001). Various cooking methods and the flavonoid content in onion. *J. nutr.Sci. vitaminology,* 47, 78-83.

Irkin, R., Korukluoglu, M. (2009). Control of Some Filamentous Fungi and Yeasts by Dehydrated Extracts. J. VerbraucherschutzL ebensmittelsicherh, 4, 3-6.

Irkin, R., Arslan, M. (2010).Effect of onion (*Allium cepa L.*) extract on microbiological quality of refrigerated beef meat. *J. Muscle Foods* , 21, 308-316.

Janoszka, B. (2010). Heterocyclic amines and azaarenes in pan-fried meat and its gravy fried without additives and in the presence of onion and garlic. *FoodChem.* 120, 463-473.

Kawamoto, E., Sakai, Y., Okamura, Y., Yamamoto, Y. (2004). Effects of boiling on the antihypertensive and antioxidant activities of onion. *J. Nutr. Sci. Vitaminol.,* 50, 171–176.

Kaymak-Ertekin, F., Gedik, A. (2005).Kinetic modelling of quality deterioration in onions during drying and storage. *J. Food Eng.*, 68, 443–453.

Keiss, H.P., Dirsch, V.M., Hartung, T., Haffner, T., Trueman, L., Auger, J., Kahane, R., Vollmar, A.M. (2003). Garlic (*Allium sativum L.*) modulates cytokine expression in lipopolysaccharide-activated human blood thereby inhibiting NF-kappaB activity. *J Nutr.* 133, 2171-2175.

Kim,D.H.,Jung, E.A.,Sihng,I.S.,Han,J.A.,Kim, T.H., Han, M.J. (1998).Intestinal bacterial metabolism of flavonoids and its relation to some biological activities. *Arch. Pharm. Res.*21, 17–23.

Kim, J.H. (1997). Anti-bacterial action of onion (*Allium cepa L.*) extracts against oral pathogenic bacteria. *J. Nihon. Univ. School Dentistry,*39, 136-141.

Kubec, R., Hrbácová, M., Musah, R.A., Velísek, J. (2004). Allium discoloration: Precursors involved in onion pinking and garlic greening. J. Agric. *Food Chem.* 52, 5089–5094.

Kumari, K., Mathew, B.C., Augusti, K.T. (1995).Antidiabetic and hypolipidemic effects of S-methyl cysteine sulfoxide isolated from *Allium cepa L. Indian J. BiochemBiophys,* 32, 49-54.

Lanzotti, V. (2006).The analysis of onion and garlic. *Journal of Chromatography* A, 1112, 3-22.

Lister, C.E., Skinner, M.A., Hunter, D.C. (2007).Fruits, vegetables and their phytochemicals for bone and joint health.Curr.*topics in nutraceutical* r., 5, 67-82.

Ly, T.N., Hazama, C., Shimoyamach, M., Ando, H.,Kato, K., Yamauchi, R. (2005).Antioxidative compounds from the outer scales of onion. *J. Agric. Food Chem.* 53, 8183–8189.

Makris, D.P.,Rossiter, J.T. (2001). Domestic processing of onion bulbs (*Allium cepa*) and asparagus spears (Asparagus officinalis): Effect on flavonol content and antioxidant status. *J. Agric. Food Chem.* 49, 3216-3222.

Manach, C., Morand, C., Crespy, V. (1998). Quercetin is recovered in human plasma as conjugated derivatives which retain antioxidant properties. *FEBS Letters*, 426,331.

Manach, C., Scalbert, A., Morand, C., Remesy, C., Jimenez, L. (2004). Polyphenols: food sources and bioavailability. *Am. J. Clinic. Nutr.* 79, 727-747.

Matheson, E.M.,Mainous, A.G.,Carnemolla, M.A. (2009).The association between onion consumption and bone density in perimenopausal and postmenopausal non-Hispanic white women 50 years and older. *Menopause-the journal of the north american menopause society*, 16, 756-759.

McDonald, I.A., Mader, J.A., Bussard, R.G.(1983). The role of rutin and quercitrin in stimulating flavonol glycosidase activity by cultured cell-free microbial preparations of human faeces and saliva. *Mutat Res*, 122, 95.

Miron, T., Rabinkov, A., Peleg, E., Rosenthal, T., Mirelman, D., Wilchek, M. (2004).Allylmercaptocaptopril: a new antihypertensive drug. *American Journal of Hypertension,* 17, 71-73.

Moncada, S., Higgs, E.A. (1991).Endogenous nitric oxide: physiology, pathology and clinical relevance. *Eur. J. Clin. Invest.* 21, 361-374.

Moon, C.H., Jung, Y.S., Kim, M.H., Lee, S.H., Baik, E.J., Park, S.W. (2000). Mechanism for antiplatelet effects of onion: araechidonic acid release inhibition, thromboxane A(2) synthase inhibition and TXA (2)/PGH (2)receptor blockade. *Prost LeukotEssent Fatty Acids,* 62, 277-283.

Morand, C., Manach, C., Crespy, V., Remesy, C. (2000). Respective bioavailability of quercetinaglycone and its glycosides in rat model.*Biofactors,* 12,169–174.

Murota, K., Shimizu, S., Chujo, H.,Moon, J., Terao, J. (2000). Efficiency of absorption and metabolic conversion of quercetin and its glucosides in human intestinal cell line Caco-2. *Arch. Biochem. Biophys.* 384, 391–397.

Naczk, M., Shahidi, F. (2006).Phenolics in cereals, fruits and vegetables: Occurrence,extraction and analysis. *Pharmaceutical and Biomedical Analysis,* 41, 1523–1542.

Nemeth, K., Piskula, M.K. (2007).Food Content, Processing, Absorption and Metabolism of Onion Flavonoids. *Critical Reviews in Food Science and Nutrition,* 47, 397-409.

Nemeth, K.,Piskula, M. K.,Takacsova, M. (2004).Effect of boiling on yellow onion quercetin (glucosides).*Czech J. Food Sci.* 22, 164–167.

Olthof, M., Hollman, P.C.H., Vree, T.B., Katan, M.B. (2000).Bioavailabilities of quercetin-3-glucoside and quercetin-4-glucoside do not differ in humans. *J. Nutr.,* 130, 1200.

Orvis, K.S., Goldman, I.L., Barak, P. (2001).Excess sulfate supply and onion-induced human antiplatelet activity.*Journal of plant nutrition and soil science-zeitschrift fur pflanzenernahrung und bodenkunde*, 164, 457-462.

Osmont, K.S.,Arnt, C.R., Goldman, I.L. (2003).Temporal aspects of onion-induced antiplatelet activity.*Plant foods for human nutrition*, 58, 27-40.

Ou, C.C., Tsao, S.M., Lin, M.C., Yin, M.C. (2003). Protective action on human LDL against oxidation and glycation by four organosulfur compounds derived from garlic. *Lipids*. 38, 219-224.

Paganga, G., Miller, N., Rice-Evans, C.A. (1999).The polyphenolic content of fruit and vegetables and their antioxidant activities. What does a serving constitute? *Free Radical Res*, 30, 153-162.

Park, S.Y., Yoo, S.S., Shim, J.H., Chin, K.B. (2008). Physicochemical properties, and antioxidant and antimicrobial effects of garlic and onion powder in fresh pork belly and loin during refrigerated storage. *J. Food Sci.* 73.

Pellegrini, N.,Miglio, C., Del Rio, D., Salvatore, S., Serafini, M.,Brighenti, F. (2009).Effect of domestic cooking methods on the total antioxidant capacity of vegetables. *Int. J. Food Sci. Nutr.* 60, 12-22.

Piskula, M. K.,Terao, J.(1998).Accumulation of (-)-epicatechin metabolites in rat plasma after oral administration and distribution of conjugation enzymes in rat tissues. *J. Nutr.* 128, 1172–1178.

Ponce, A.G.,Roura, S.I., del Valle, C.E., Moreira, M.R. (2008).Antimicrobial and antioxidant activities of edible coatings enriched with natural plant extracts: In vitro and in vivo studies. Postharvest Biol. *Technol.* 49, 294-300.

Prasad, K., Laxdal, V.A., Yu, M., Raney, B.L. (1995).Antioxidant activity of allicin, an active principle in garlic.*Mol. Cell Biochem.* 148, 183-189.

Price, K.R., Bacon, J.R., Rhodes, M.J.C.(1997). Effect of storage and domestic processing on the content and composition of flavonolglucosides in onion (*Allium cepa). J. Agric. Food Chem*,45,938–942.

Price, K.R., Rhodes, M.J.C. (1997). Analysis of the major flavonol glycosides present in four varieties of onion and changes in composition resulting from autolysis. *J. Sci .Food Agric*, 74, 331.

Price, S.F., Breen, P.J., Valladao, M., Watson, B.T. (1995). Cluster sun exposure and quercetin in pinot noir grapes and wine. *Am. J. Enol. Vitic.* 46, 187-194.

Ramadan, M. F.,Selim, M. M. (2009).Antimicrobical and antivirial impact of novel quercetin-enriched lecithin. *Journal of Food Biochemistry,* 33, 557–571.

Rodrigues, A.S., Pérez-Gregorio, M.R., García-Falcón, M.S.,Simal-Gándara, J. (2009).Effect of curing and cooking on flavonols and anthocyanins in traditional varieties of onion bulbs. *Food Res. Int.*, 42, 1331-1336.

Rohn, S., Buchner, N., Driemel, G., Rauser, M.,Kroh, L.W. (2007).Thermal degradation of onion quercetinglucosides under roasting conditions. *J. Agric. Food Chem.* 55, 1568-1573.

Roldan, E., Sanchez-Moreno, C., de Ancos, B., Cano, M.P. (2008). Characterisation of onion (*Allium cepa L.*) by-products as food ingredients with antioxidant and antibrowning properties. *Food Chem.* 108, 907-916.

Rose, P., Whiteman, M., Moore, P.K., Zhu, Y.Z.,. (2005). Bioactive S-alk(en)yl cysteine sulfoxide metabolites in the genus Allium: the chemistry of potential therapeutic agents. *Nat. Prod. Rep.,* 22, 351–368.

Rose, P., Whiteman, M., Moore, P.K., Zhu, Y.Z. (2005). Bioactive S-alk(en)yl cysteine sulfoxide metabolites in the genus Allium: the chemistry of potential therapeutic agents. *Natural product reports,* 22, 351-368.

Santas, J., Almajano, M.P., Carbó, R.(2010).Antimicrobial and antioxidant activity of crude onion (*Allium cepa, L.*) extracts. *International Journal of Food Science and Technology,* 45, 403–409.

Santas, J.,Almajano, M.P., Carbó, R.(2010).Onion A natural alternative to artificial food preservatives.*AgroFood industry hi-tech*, 21, 44-46.

Santas, J., Carbó, R., Gordon, M.H.,Almajano, M.P. (2008).Comparison of the antioxidant activity of two Spanish onion varieties. Food Chem. ; *Food Chem.* 107, 1210-1216.

Sesink, A.L., Faassen-Peters, M., Hollman, P.C.,Arts, I.C. (2004). The type of sugar moiety is a major determinant of the small intestinal uptake and subsequent biliary excretion of dietary quercetin glycosides. *Br. J Nutr.*,91, 841-847.

Siess, M.H., Le Bon, A.M., Canivenc-Lavier, M.C., Suschetet, M. (1997). Modification of hepatic drug metabolizing enzymes in rats treated with alkyl sulfides. *Cancer Letters*, 120, 195-201.

Sivam, G.P. (2001).Protection against helicobacter pylori and other bacterial infections by garlic. Journal of nutrition, 31, 1106S-1108S.

Škerget, M., Majheniě, L., Bezjak, M.,Knez, Ž. (2009). Antioxidant, radical scavenging and antimicrobial activities of red onion (*Allium cepa L.*) skin and edible part extracts. *ChemBiochemEng Q,* 23, 435-444.

Slimestad, R., Fossen, T.,Vågen, I.M. (2007). Onions: A source of unique dietary flavonoids. J. Agric. Food Chem., 55, 10067-10080.

Sosulski, F., Krygier, K., Hogge, L. (1982).Free, esterified, and insoluble-bound phenolic acids. Composition of phenolic acids in cereal and potato flours. *J. Agric. Food Chem,* 30, 337–340.

Spanos, G.A., Wrolstad, R.E.(1992). Phenolics of apple, pear, and white grape juice and their changes with processing and storage—a review. *J. Agric. Food Chem, 40*, 1478–1487.

Srivastava, K.C. (1986).Evidence for the mechanism by which garlic inhibits platelet aggregation.*Prostaglandins Leukot. Med.* 22, 313-21.

Stewart, A.J.,Bozonnet, S., Mullen, W., Jenkins, G., Lean, M., Alan Crozier, A. (2000).Occurrence of Flavonols in Tomatoes and Tomato-Based Products. *J. Agric. Food Chem.* 48, 2663–2669.

Stratil, P.,Klejdus, B., Kuban, V. (2006).Determination of total content of phenolic compounds and their antioxidant activity in vegetables - Evaluation of spectrophotometric methods. *J. Agric. Food Chem,* 54, 607-616.

Suh, H.J., Leeb, J.M., Chob, J.S., Kima, Y.S., Chunga, S.H. (1999). Radical scavenging compounds in onion skin. *Food Res. Int.* 32, 659–664.

Takahama, U., Takayuki,T., Hirota, S. (2001). Phenolic components of brown scales of onion bulbs produce hydrogen peroxide by autooxidation. *J. Plant Res.* 114, 395–402.

Takahma, U., Hirota, S. (2000).Deglucosidation of quercetinglucosides to the aglycone and formation of antifungal agents by peroxidase-dependent oxidation of quercetin on browning of onion scales. *Plant Cell Physiol.* 41, 1021–1029.

Takenaka, M., Nanayama, K., Ohnuki, I., Udagawa, M., Sanada, E., Isobe, S. (2004). Cooking loss of major onion antioxidants and the comparison of onion soups prepared in different ways. *Food Sci. Technol. Res.* 10, 405–409.

Tang, X., Cronin, D.A. (2007).The effects of brined onion extracts on lipid oxidation and sensory quality in refrigerated cooked turkey breast rolls during storage. .., 100, 712-718.

van der Sluis, A.A., Dekker, M., de Jager, A., Jongen, W.M.(2001). Activity and concentration of polyphenolic antioxidants in apple: effect of cultivar, harvest year, and storage conditions. *J. Agric. Food Chem*, 49, 3606–3613.

Vazquez-Prieto, M., Roberto, M. (2010). *MiatelloOrganosulfur compounds and cardiovascular disease Molecular Aspects of Medicine,* 31, 540–545.

Vidyashankar, S.,Sambaiah, K.,Srinivasan, K. (2010).Regression of preestablished cholesterol gallstones by dietary garlic and onion in experimental mice.*Metabolism-clinical and experimental*, 59, 1402-1412.

Walle, T., Otake, Y., Walle, U.K., Wilson, F.A.(2000). Quercetinglucosides are completely hydrolyzed in ileostomy patients before absorption. *J. Nutr* ,130,2658.

Xiao, H., Parkin, K.L. (2002). Antioxidant functions of selected allium thiosulfinates and S-alk(en)yl-L-cysteine sulfoxides. *J. Agric. Food Chem.* 50, 2488-2493.

Xu, H.X., Lee, S.F. (2001).Activity of plant flavonoids against antibiotic-resistant bacteria.*Phytotherapy Research*, 15, 39-43.

Yang, J., Meyers, K.J., Van, D.H., Rui, H.L. (2004).Varietal differences in phenolic content and antioxidant and antiproliferative activities of onions. *Journal of Agricultural and Food Chemistry,* 52, 6787–6793.

Yin, M.C., Tsao, S.M. (1999).Inhibitory effect of seven Allium plants upon three Aspergillus species.*Int. J. Food Microbiol*, 49, 49-56.

You, W.C., Zhang, L., Gail, M.H., Ma, J.L., Chang, Y.S., Blot, W.J., Li, J.Y., Zhao, C.L., Liu, W.D., Li, H.Q., Hu, Y.R., Bravo, J.C., Correa, P., Xu, G.W., Fraumeni, J.F. Jr. (1998).Helicobacter pylori infection, garlic intake and precancerous lesions in a Chinese population at low risk of gastric cancer.*Int. J. Epidemiol.* 27, 941-944.

Zohri, A.N., Abdel-Gawad, K., Saber, S.(1995).Antibacterial, antidermatophytic and antioxigenic activities of onion (*Allium cepa L.*) oil.*Microbiol Res.* 150, 167-172.

Zubay, G.(1986). Biochemistry.California: Addison-Wesley.

In: Onion Consumption and Health
Editors: C. B. Aguirre et al.
ISBN: 978-1-62100-836-1

Chapter 4

ONION BIOACTIVE COMPOUNDS AND HEALTH EFFECTS

Eduvigis Roldán-Marín, Begoña de Ancos[a], M. Pilar Cano[b], Concepción Sánchez-Moreno[a,*]

[a]Department of Food Characterization, Quality and Safety, Institute of Food Science, Technology and Nutrition (ICTAN), Spanish National Research Council (CSIC), Madrid, Spain

[b]Department of Food Biotechnology and Microbiology, Institute of Food Science Research (CIAL), Spanish National Research Council (CSIC)-UAM, Madrid, Spain

ABSTRACT

Nowadays, there is a constant and increasing social concern about the food we daily eat and particularly about vegetables. Moreover, there is a worldwide great demand of minimally processed food or food ingredients with functional properties as a consequence of the changing healthy and environmental sustainable living that is becoming to spanning almost the totality of the world's cultures.

[*] Corresponding author: Concepción Sánchez-Moreno. Department of Food Characterization, Quality and Safety, Institute of Food Science, Technology and Nutrition (ICTAN), Spanish National Research Council (CSIC), José Antonio Novais 10, Ciudad Universitaria. E-28040 Madrid, Spain; E-mail: csanchezm@ictan.csic.es

The Mediterranean diet which traditionally includes generous amounts of fruits and vegetables has been associated with health benefits such as lower risks for cardiovascular disease (CVD) and cancer. There is epidemiologic evidence that frequent consumption of *Allium* vegetables such as garlic, scallions, onions, chives, and leeks is protective against cancer. Therefore, *Allium* vegetables should be considered to be an important component of a healthy, cancer and CVD resistant diet.

We consider that it is needed to study in depth not only the technological but also the biological and nutritional properties of vegetables products in order to better understand their role in our daily diet and lifestyle. In this chapter onion has been chosen as a model of an *Allium* vegetable due to its good technological and nutritional properties reported and to its presence in the daily diet worldwide. Onion is a good source of bioactive compounds such as flavonoids, organosulfur compounds (OSCs), fructans and fructooligosaccharides (FOS). Onion has been described to have several health benefits related to its antioxidant, anticarcinogenic, hypolipidemic, hypoglycaemic, or antiaggreagatory effects.

Introduction

Onions (*Allium cepa* L.) are bulbous vegetables from the *Liliaceae* family, important in terms of domestic consumption and export.

Onions are grown in at least 175 countries, two of the leading producers are China and India. According to the Food and Agriculture Organization (FAO) of the United Nations, onions are the second most important horticultural crop worldwide, after tomatoes, with a world production in 2009 around 73 million tonnes. Over the past 10 years, onion production has increased by more than 25% (FAO, 2009)

Onions are grown mainly as food materials. They are highly valued for their flavour and for their nutritional value. Onion bulb which may be red, white or yellow in colour, is consumed in its tender state, raw, ripe, pickled or in form of powder. The bulbs are boiled and used in soups and stews, fried or eaten raw. They are also preserved in the form of pickles. Onion leaves are also used in salads and soups (Paltrinieri *et al.*, 1997).

ONION TECHNOLOGICAL PROPERTIES

From a technological point of view onion has interesting properties. Among them, onions have been reported to have antibrowning, antioxidant, and antimicrobial properties. These onion technological properties represent an added value to onion that would give a new challenge for the food industry nowadays.

Onion Antibrowning Properties

The browning reaction is a widespread phenomenon in fruits and vegetables, resulting from mechanical or physiological injury during post-harvest storage or processing. It is a major factor contributing to quality loss in foods and beverages. The polyphenol oxidase (PPO) enzyme mainly caused the enzymatic oxidation of endogenous phenols into quinones, which then polymerize into brown products. Thus, polyphenolic compounds and PPO are directly responsible for the enzymatic browning.

The use of sulfiting agents is the most widespread chemical approach for controlling browning. However, consumer awareness of the risks associated with sulfites and increased regulatory scrutiny have created the need for substitutes (Iyengar & McEvily, 1992). Among the compounds that have been shown to inhibit the PPO activity are sulfites, ascorbic acid and its derivatives, and thiol compounds such as cysteine (Martínez & Whitaker, 1995; Negishi & Ozawa, 2000; Ding *et al.*, 2002; Jang *et al.*, 2002).

Onion has been reported to inhibit PPO of fruits such as pear or banana. Kim *et al.* (2005) reported that thiol compounds in onion might be the active components responsible for the inhibitory effect of onion extract. When the onion extract was dialyzed, the inhibitory effect against pear PPO was completely eliminated, suggesting that the low molecular compounds were responsible of the pear PPO inhibitory effects. Moreover, when onion extracts were heated the pear PPO inhibition was more efficient. Lee (2007) also found that heat treated onion extracts inhibited banana PPO. Furthermore, it was shown that Maillard reaction products significantly inhibited banana PPO as well as the addition of various antibrowning agents.

Onion Antioxidant Properties

From a nutritional point of view, it has been shown that *Allium* species may help to prevent tumour promotion, cardiovascular diseases and aging; all processes that are associated with free radicals (Stajner *et al.*, 2006).

In terms of onion *in vitro* antioxidant activity, the assessment of the antioxidant activities and comparison with previously reported findings show that the antioxidant activity of 5 portions of onion is equivalent to 1 glass (150 mL) red wine, 12 glasses white wine, 2 cups of tea, 4 apples, 5.5 portions egg plant, 3.5 glasses of blackcurrant juice, 3.5 (500 mL) glasses of beer, 7 glasses of orange juice, and 20 glasses of apple juice (long life) (Paganga *et al.*, 1999).

Onions had clearly higher radical scavenging activities than garlic, red onion being more active than yellow onion. Quercetin and its derivates together with the anthocyans, have been widely reported to have antioxidant properties. Flavonoid compounds and particularly those having o-dihydroxy substituent in the B-ring were shown to be effective antioxidants (Ly *et al.*, 2005). It was demonstrated that the skin extracts of onion possessed the highest antioxidant activities.

Levels of quercetin, $DPPH^{\bullet}$ radical scavenging activity, and antioxidant capacity increase from onion core to skin (Nuutila *et al.*, 2002; Nuutila *et al.* 2003; Ly *et al.*, 2005; Kim & Kim, 2006; Benítez *et al.*, 2011). Onion phenolic and sulfur compounds are among the onion bioactive compounds involved in the assessment of onion antioxidant properties (Benkeblia *et al.*, 2005).

Moreover, onion *in vitro* antioxidant activity depends not only on the onion varieties or cultivars (Yang *et al.*, 2004; Santas *et al.*, 2008) but also on the processing or on the different heat treatment applied to onions (Yang *et al.*, 2004; Moreno *et al.*, 2006; Woo *et al.*, 2007).

In terms of onion *in vivo* antioxidant activity, different studies carried out in rats or in humans assessed that onion has positive health effects acting as an antioxidant vegetable helpful in the prevention of some nowadays prevalent diseases such as cardiovascular disease (CVD). Several onion fed rat studies corroborated those onion antioxidant effects (Azuma *et al.*, 2007; Park *et al.*, 2007; Slobodianik *et al.*, 2007; Roldán-Marín *et al.*, 2009a). The same antioxidants effects have been shown in human studies including onion in the diet (McAnlis *et al.*, 1999; Boyle *et al.*, 2000).

Onion Antimicrobial Properties

Onions have few applications of *Allium* vegetables as natural food preservatives, in spite of numerous studies on antimicrobial activity of these vegetables. Relative instability of the antimicrobial compounds and the strong odour of their mother plants seem to limit the use of them as practical food preservatives (Kyung & Lee, 2001).

A recent study showed that the addition of garlic and onion powders in enhanced meats had an antioxidant activity as effective as that of sodium ascorbate and also an antimicrobial effect to inhibit the growth of total bacteria and bacteria from the *Enterobacteriaceae* family (Park *et al.*, 2008).

Different onion extracts and oils have proven antimicrobial and antioxidant properties which are interesting from a technological and nutritional point of view (Irkin & Korukluoglu, 2007; Choi *et al.*, 2008). Organosulfur compounds (OSCs) and phenolic compounds have been reported to be involved in the onion antimicrobial activity (Takahama & Hirota, 2000; Griffiths *et al.*, 2002; Kim *et al.*, 2004a; Kim *et al.*, 2004b). Other compounds such as the peptide allicepin have antifungal properties and have been isolated from onion bulbs (Wang & Ng, 2004).

Onion By-products

More than 450000 tonnes of onion wastes are produced annually in Europe, mainly in the United Kingdom, Holland, and Spain. There is considerable industrial pressure to come up with a way to convert the waste into useful products. This attempt would reduce the environmental impact of onion waste disposal by converting waste streams into useful products resulting in low-waste food production (Waldron, 1999; Waldron, 2001). By-products are promising sources of compounds which may be used because of their favourable technological or nutritional properties (Schieber *et al.*, 2001). The exploitation of by-products of onion processing as a source of functional compounds and their application in food is a promising field which requires interdisciplinary research of food technologists, food chemists, nutritionists and toxicologists. Active participation of the food and allied industries with respect to sustainable production and waste management is required. Roldán *et al.* (2008) demonstrated that processing and stabilizing onion wastes (residues and surpluses of onion) could represent both advantages: a solution for the environmental problem derived from the great onion wastes disposal and the obtaining of stabilized onion by-products as natural antioxidant food ingredients.

High-pressure Processed Onion Products

High-pressure processing (HPP) is an industrially tested technology that offers a natural alternative for the processing of a wide range of different food products. This waste-free and environmentally friendly technology can achieve the food safety of heat pasteurization whilst meeting consumer demand for fresher-tasting minimally processed foods (Norton & Sun, 2008).

Processing fresh onion with the nonthermal technology high hydrostatic pressure combined with low temperatures might have additional advantages than processing fresh onion with conventional thermal technologies that apply high temperatures. Therefore, the high-pressure processing technology could represent an alternative to some of these traditionally used thermal technologies.

Processing fresh onion with high-pressures (400 MPa) and low temperatures (5 °C) could positively enhance onion bioactive compounds extractabilities leading to high-pressure (HP)-processed onion products that have higher bioactive compound content respect to fresh onion. In the same way, this onion processing at 400 MPa/5 °C maintained the antioxidant activity of fresh onion. Therefore, HP-processed onion products offer an interesting nutritional added value that maintains fresh onion antioxidant capacity (Roldán-Marín *et al.*, 2009b).

Onion Biological and Nutritional Properties

Onion Nutritional Content

Onion (*Allium cepa* L.) has an approximately 90% content of water. Onion is low in calories and has a moderately high content of dietary fibre and sugars (Table 1).

In terms of vitamin and mineral content onion has low sodium content and a high content of vitamin B_6, folic acid, calcium, magnesium, phosphorus and potassium. By contrast, onion has low content of lipids and among the amino acid content only arginine and glutamic acid are remarkable.

Table 1. Raw onion (*Allium cepa* L.) proximates

Proximates	Unit	Mean ± SD
Water	g	89.11 ± 0.248
Energy	kcal	40 ± 0
Energy	kJ	166 ± 0
Protein (N x 6.25)	g	1.10 ± 0.036
Total lipid (fat)	g	0.10 ± 0.005
Ash	g	0.35 ± 0.003
Carbohydrate, by difference	g	9.34 ± 0
Fibre, total dietary	g	1.7 ± 0.048
Sugars, total	g	4.24 ± 0
Sucrose	g	0.99 ± 0.050
Glucose (dextrose)	g	1.97 ± 0.054
Fructose	g	1.29 ± 0.052

Amount in 100 grams of edible portion USDA (2008).

Onion Bioactive Compounds

Onion added into different food products as a food ingredient made these products richer in bioactive compounds such as flavonoids, dietary fibre, and organosulfur compounds (OSCs). These onion bioactive compounds have potential beneficial health effects. Moreover, some of the onion bioactive compounds are also interesting from a technological point of view due to their properties as preservatives agents, OSCs due to their antioxidant and antibrowning properties or dietary fibre due to possibly texture enhancer properties.

Flavonoids, Quercetin and Quercetin Glucosides

Onion is among the richest source of dietary flavonoids and contributes to a large extent to the overall intake of flavonoids. Two flavonoid subgroups are found in onion, the anthocyanins, which impart a red/purple colour to some varieties and the flavonols such as quercetin and its derivatives responsible for the yellow and brown skins of many other varieties (Griffiths *et al.*, 2002). Flavonols are the predominant pigments of onions. Only compounds belonging to the flavonols, the anthocyanins, and the dihydroflavonols have been reported to occur in onion bulbs.

At least 25 different flavonols have been characterized, and quercetin derivatives are the most important ones in all onion cultivars. Their glycosyl moieties are almost exclusively glucose, which is mainly attached to the 4', 3, and/or 7-positions of the aglycones. Analogous derivatives of kaempferol and isorhamnetin have been identified as minor onion pigments (Slimestad *et al.*, 2007).

Quercetin 3,4'-diglucoside (Q_{DG}) and quercetin 4'-glucoside (Q_{MG}) are in most cases reported as the main onion flavonols in recent literature (Caridi *et al.*, 2007). These glucosides of quercetin represent about the 90% of the overall contents in different *Allium* species (Bonaccorsi *et al.*, 2008). Significant differences in the levels and ratios of the two compounds were seen between red ('Redwing'; Q_{DG} 191 mg/100 g dry weight-dw and Q_{MG} 85 mg/100 g dw), brown ('Cream Gold', Q_{DG} 153 mg/100 g dw, Q_{MG} 58 mg/100 g dw), and white onion varieties ('Spanish white'; Q_{DG} < 1 mg/100 g dw, Q_{MG} < 1 mg/100 g dw) (Caridi *et al.*, 2007).

Distribution of quercetin and its glycosides within the onion bulb changes during onion processing, different `cooking´ methods, and exposure to fluorescent light (Ewald *et al.*, 1999; Mogren *et al.*, 2007; Nemeth & Piskula, 2007; Lee *et al.*, 2008b; Mogren *et al.*, 2008). After ingestion quercetin and its glycosides undergoes extensive metabolism and microbial action resulting in its altered or degraded structure; therefore, most of the effects shown in *in vitro* experiments with the pure compound cannot be directly extrapolated to *in vivo* systems (Nemeth & Piskula, 2007).

Absorption and metabolism of the flavonol quercetin and its glycosides have been described by Aherne & O'Brien (2002). Quercetin glycosides, the mostly present form in onions, are converted to the respective aglycones in the large intestine by the glycosidase activity of intestinal bacteria and absorbed. Several studies have demonstrated that quercetin glycosides are absorbed more efficiently than quercetin aglycone irrespective of the position of their glucose moiety (Hollman *et al.*, 1995; Moon *et al.*, 2000). After being absorbed, quercetin is metabolized and excreted (Day *et al.*, 2001). Mullen *et al.* (2006) found five metabolites in quantifiable amounts in human plasma after onion ingestion (quercetin 3-glucuronide, quercetin 3'-sulfate, isorhamnetin 3-glucuronide, a quercetin diglucuronide and a quercetin glucuronide sulfate). They also reported that total urinary excretion of quercetin metabolites was 12.9 mmol, corresponding to 4.7% of intake.

In general, quercetin has been investigated for their widespread health benefits which have generally been ascribed to its combination of antioxidant and antiinflammatory activities (Davis *et al.*, 2009). Quercetin beneficial health effects include protection against various diseases such as osteoporosis, certain forms of cancer, pulmonary, and CVD but also against aging. Especially the ability of quercetin to scavenge highly reactive species such as peroxynitrite and the hydroxyl radical is suggested to be involved in these possible beneficial health effects (Boots *et al.*, 2008).

Organosulfur Compounds (OSCs)

Allium plants contain high concentrations of alk(en)yl cysteine sulfoxides (ACSOs). Among them, onions, shallots, and leeks contain the methyl, propyl, and prop-1-enyl cysteine sulfoxides.

When the onion tissue is disrupted by cutting, crushing or chewing, the ACSOs are enzymatically degraded by the enzyme alliinase to iminopropionic acid and alk(en)yl cysteine sulfenic acids.

The iminopropionic acid spontaneously hydrolyses to form ammonia and pyruvic acid. The sulfenic acids decompose spontaneously. Methyl and propyl sulfenic acids yield mainly thiosulfinates, while prop-1-enyl sulfenic acid forms both, the corresponding thiosulfinate and thiopropanal S-oxide, the onion lachrymatory factor (Block *et al.*, 1993).

Thiosulfinates themselves are unstable, particularly on heating, and break down to a complex mixture of compounds, in which mono-, di-, tri- and tetra-sulfides predominate (Munday & Munday, 2004; Rose *et al.*, 2005).

Dipropyl disulfide, dipropyl trisulfide, and propenyl disulfides are the major constituents of onion volatiles, although many others compounds have been identified, including dipropyl sulfide and dipropenyl sulfide (Munday & Munday, 2001).

Defining the mechanism by which organosulfur compounds (OSCs) derived from *Allium* vegetables inhibit cancer cell growth has been the topic of intense research in the last two decades. Some *Allium* vegetable constituents have also entered clinical trials to assess their safety and anticancer efficacy (Powolny & Singh, 2008). OSCs have been shown to exert diverse biological effects such as: induction of carcinogen detoxification, inhibition of tumor cell proliferation, antimicrobial effect, free radical scavenging, inhibition of DNA adduct formation, induction of cell cycle arrest, and induction of apoptosis. It has been suggested that these compounds act as chemopreventive agents through a combination of above mechanisms (Moriarty *et al.*, 2007).

It has also been proposed that the chemoprotective effect of onion sulfides is due, at least in part, to the ability of these sulfides to increase tissue activities of phase II detoxification enzymes (Guyonnet *et al.*, 2001; Teyssier *et al.*, 2001; Vázquez-Prieto & Miatello, 2010).

Taking into account the relationship between structure and enzyme inducing ability by OSCs it can be drawn the importance of unsaturation in the alkyl chain, showing prop-1-enyl derivatives a higher level of induction than the propyl in many tissues. The importance of the number of sulfur atoms in the molecule was also important; the general rule could be that the higher activity is found with increasing number of sulfur atoms, although the direction of the effect was different in the prop-1-enyl sulfide which was stronger inducer than the disulfide (Munday & Munday, 2004).

Fructans and Fructooligosaccharides (FOS)

Bulb dry matter content is an important quality parameter of onion, also important in the onion dehydration industry because it directly relates to the energy needed for drying. About 65 to 80% of bulb dry matter consists of non-structural carbohydrates (Kahane *et al.*, 2001; Benkeblia *et al.*, 2005).

The onion predominant non-structural carbohydrates are glucose, fructose, sucrose and low-molecular-weight fructans, while starch and raffinose are absent. Fructans, also known as fructooligosaccharides (FOS), are polyfructosyl sucroses of varying molecular size that constitute the main carbohydrate reserve of onion. Fructans accumulate during bulbing and are then catabolized during regrowth and sprout development of the bulbs (Benkeblia *et al.*, 2005).

It is generally accepted that FOS is a common name only for fructose oligomers that are mainly composed of kestose (GF_2), nystose (GF_3) and fructofuranosylnystose (GF_4), in which fructosyl units (F) are bound by ß-linkage at the position of sucrose (glucose+fructose-GF) respectively. There is a clear predominance of kestose (GF_2) in every onion tissue and no occurrence of highly polymerized fructans. The tissues richest in fructans are the fleshy layers, so that the outer two fleshy layers turn out to be the best onion by-product as a possible fructan source (Jaime *et al.*, 2001a; Jaime *et al.*, 2001b; Jaime *et al.*, 2002).

The fructan degree of polymerization level in onion is mostly in between 3 and 15. Short chain fructans, with a degree of polymerization (DP) less than 5, are potentially used as natural low-calorie sweeteners. Onion bulbs with fructans of a high DP may be used for lipid replacement with consequential health benefits (Kahane *et al.*, 2001).

Onion showed a better soluble/insoluble dietary fibre (SDF:IDF) ratio than other vegetables that will be connected with different metabolic and physiological effects SDF increases the viscosity of the stomach contents, thereby allowing down-mixing and absorption of nutrients, whereas IDF reduces intestinal transit time and increases the bulk of the food mass (Jaime *et al.*, 2002). Fructans could act as osmoregulators due to their solubility in water inside the vacuole.

Fructans act stimulating the growth of specific microorganisms in the colon (*Bifidobacteria* and *Lactobacilli*) with a general positive health effect, including on colonic inflammation (Ernst & Feldheim, 2000; Lara-Villoslada *et al.*, 2006; Roberfroid, 2007). Administration of FOS significantly lowered fasting glycemia and total cholesterol, increasing the intestinal absorption and bone density of calcium and magnesium.

Other Onion Bioactive Compounds

Recent literature reported that several interesting novel compounds have been isolated from onion. Among them, saponins and peptides have been isolated and studied for their potentially beneficial health effects.

5-hydroxy-3-methyl-4-propylsulfanyl-5H-furan-2-one, and four others compounds, were isolated and confirmed to be quinone reductase and glutathione *S*-transferase inducers *in vitro* (Xiao & Parkin, 2007), therefore they could act as chemopreventive agents. This warrants further research to isolate and identify more agents for their potential for phase II enzyme induction *in vitro* and *in vivo*.

Several research reports have demonstrated antifungal, antitumor, cytotoxicity, blood coagulability, antispasmodic and cholesterol-lowering effects of saponins isolated from onion and garlic (Lanzotti, 2006).

Four furostanol saponins, two of which were new compounds, named ceparoside A and ceparoside B were isolated from the seeds of *Allium cepa* L. (Yuan *et a*l., 2008). Other new saponins were found years earlier by Corea *et al.* (2005), they were reported to possess antispasmodic activity in the guinea pig isolated ileum; such an effect might contribute to explain the traditional use of onion in the treatment of disturbances of the gastrointestinal tract.

Recently, it was also reported that an onion gamma-glutamyl peptide from onion (Wetli *et al.*, 2004) inhibits the development and activity of osteoclasts *in vitro* (Langos *et al.*, 2007).

Onion Health Effects

Allium vegetables health properties have been supported by numerous *in vitro*, *in vivo*, and *ex-vivo* studies. Particularly, onion has been described to have several health benefits related to its antioxidant, anticarcinogenic, hypolipidemic, hypoglycaemic, or antiaggreagatory effects.

It has been described that a diet rich in *Allium* vegetables, including onion, would lead to several and different health benefits that could be helpful in the prevention of two of the more relevant and prevalent diseases nowadays such as cancer or CVD.

Onion and Cancer

The association between consumption of *Allium* vegetables and risk for cancer has been assessed in several epidemiologic studies, mainly case-control (Bianchini & Vainio, 2001; Galeone *et al.*, 2006).

In general, these studies are more consistent in reporting a protective effect of onion in gastric cancer. However, onion consumption has been also consistently related with a decreased colorectal cancer risk.

In addition, onion consumption was reported to decrease the risk for the cancer of the lung (Sankaranarayanan *et al.*, 1994) and of the brain (Hu *et al.*, 1999) in case-control studies.

Onion consumption was significantly inversely correlated with the risk of the stomach cancer. Recently, Gónzalez *et al.* (2006) observed in the European Prospective Investigation into Cancer and Nutrition (EPIC-EURGAST) a probable protective effect of total vegetables and *Allium* vegetables intake on the intestinal type of gastric cancer (González *et al.*, 2006). Most of the case-control studies concerning onion were conducted in China (You *et al.*, 1989; Gao *et al.*, 1999) and several of them in Asia and Europe (Boeing *et al.*, 1991; González *et al.*, 1991; Tuyns *et al.*, 1992; Hansson *et al.*, 1993).

Onion and Cardiovascular Disease (CVD) Prevention. Antiaggregatory and Hypolipidemic Effects

The study by Galeone *et al.* (2009), the first from Mediterranean countries, suggests that a diet rich in onions may have a favourable effect on the risk of acute myocardial infarction; therefore these vegetables could be useful in a CVD preventive diet.

Alterations in lipid profiles, diabetes, hypertension, and obesity are risk factors conventionally associated to the early appearance of CVD. Onion has been described to have hypolipidemic, hypoglycaemic, and antithrombotic effects and therefore could be useful in the CVD prevention.

Focusing on onion lipid lowering effects, this vegetable has been reported to exert moderately hypolipidemic effects in experimental animal (Roldán-Marín *et al.*, 2010) as healthy pigs fed a high fat diet and consequently potentially reduce risk indices of CVD and obesity (Ostrowska *et al.*, 2004; Gabler *et al.*, 2006).

Among bioactive compounds involved in onion hypolipidemic effects, quercetin has shown to have the ability to reduce serum cholesterol levels and arteriosclerosis severity (Glässer *et al.*, 2002). A recent study by Kumari & Augusti, 2007 also proclaimed for the lipid lowering action of the S-methyl cysteine sulfoxide isolated from *Allium cepa* L.

Onion has also been reported to have hypoglycaemic effects (Srinivasan, 2005). It was inferred that this beneficial ameliorating influence of dietary onion on diabetic nephropathy may be mediated through onion's ability to lower blood cholesterol levels and to reduce lipid peroxidation, dietary onion caused significant beneficial modulation of the progression of renal lesions in diabetic rats (Babu & Srinivasan, 1999). Other rat studies have assessed onion hypoglycaemic effects (El-Demerdash *et al.*, 2005). Recently, a study by Lee *et al.* (2008a) showed that onion skin was effective in controlling hyperglycemia in animal models of type 2 diabetes mellitus, at least in part by inhibiting alpha-glucosidase activity.

Thrombosis complications play a major role in CVD. Onion inhibits platelet aggregation *in vitro* and *in vivo* (Ali *et al.*, 1999; Ali *et al.*, 2000; Briggs *et al.*, 2001; Jung *et al.*, 2002; Hubbard *et al.*, 2006; Gorinstein *et al.*, 2010). The mechanism by which onion exerts its antithrombotic effect has been shown to involve the inhibition of thromboxane A2 formation, potent inducer of platelet aggregation (Moon *et al.*, 2000).

The antiplatelet activity observed in onion is influenced by genotype, environmental factors and genotypically determined sulfur content of the bulb (Goldman *et al.*, 1996; Sance *et al.*, 2008) having onion α-sulfinil-disulfides (cepaenes) a demonstrated antithrombotic activity (Block *et al.*, 1997).

Other Onion Health-promoting Health Effects

Onion consumption has also been reported to be involved in the bone metabolism and in the behaviour as a possible antidepressant agent.

A recent study by Matheson *et al.* (2009) reported that onion consumption seems to have a beneficial effect on bone density in perimenopausal and postmenopausal women. Furthermore, older women who consume onions most frequently may decrease their risk of hip fracture by more than 20% *versus* those who never consume onions. Prevention of low bone mass is important to reduce the incidence of osteoporotic fractures. Onion retains its bone resorption inhibitory activity in the rat when added to a vegetarian diet (Mühlbauer *et al.*, 2002).

Another recent study by Sakakibara *et al.* (2008) suggests that onion exerted antidepressant-like activity in a behavioural model that acted independently of the hypothalamic-pituitary-adrenal axis.

ACKNOWLEDGMENTS

This work has been financially supported by the Spanish Ministry of Science and Innovation [AGL2010-15910 (subprogram ALI); AGL2008-04798/C02-01/ALI; Program Consolider-Ingenio 2010, FUN-*C*-FOOD, CSD2007-00063] and the Comunidad de Madrid (ALIBIRD, S2009/AGR-1469).

REFERENCES

Aherne SA, O'Brien NM (2002). Dietary flavonols: Chemistry, food content, and metabolism. *Nutrition,* 18:75-81.

Ali M, Bordia T, Mustafa T (1999). Effect of raw *versus* boiled aqueous extract of garlic and onion on platelet aggregation. Prostaglandins, *Leukotrienes and Essential Fatty Acids,* 60:43-47.

Ali M, Thomson M, Afzal M (2000). Garlic and onions: their effect on eicosanoid metabolism and its clinical relevance. *Prostaglandins, Leukotrienes and Essential Fatty Acids,* 62:55-73.

Azuma K, Minami Y, Ippoushi K, Terao J (2007). Lowering effects of onion intake on oxidative stress biomarkers in streptozotocin-induced diabetic rats. *Journal of Clinical Biochemistry and Nutrition,* 40:131-140.

Babu PS, Srinivasan K (1999). Renal lesions in streptozotocin-induced diabetic rats maintained on onion and capsaicin containing diets. *Journal of Nutritional Biochemistry, 10*:477-483.

Benítez V, Mollá E, Martín-Cabrejas MA, Aguilera Y, López-Andréu FJ, Cools K, Terry LA, Esteban RM (2011). Characterization of industrial onion wastes (*Allium cepa* L.): Dietary fibre and bioactive compounds. *Plant Foods for Human Nutrition* 66:48-57.

Benkeblia N, Onodera S, Shiomi N (2005). Variation in 1-fructo-exohydrolase (1-FEH) and 1-kestose-hydrolysing (1-KH) activities and fructo-oligosaccharide (FOS) status in onion bulbs. Influence of temperature and storage time. *Journal of the Science of Food and Agriculture,* 85:227-234.

Bianchini F, Vainio H (2001). *Allium* vegetables and organosulfur compounds: Do they help prevent cancer? *Environmental Health Perspectives,* 109:893-902.

Block E, Gulati H, Putman D, Sha D, You N, Zhao S-H (1997). *Allium* chemistry: Synthesis of 1-[alk(en)ylsulfinyl]propyl alk(en)yl disulfides (cepaenes), antithrombotic flavorants from homogenates of onion (*Allium cepa*). *Journal of Agricultural and Food Chemistry,* 45:4414-4422.

Block E, Naganathan S, Putman D, Zhao S-H (1993). Organosulfur chemistry of garlic and onion: Recent results. *Pure and Applied Chemistry,* 65:625-632.

Boeing H, Jedrychowski W, Wahrendorf J, Popiela T, Tobiasz-Adamczyk B, Kulig A (1991). Dietary risk-factors in intestinal and diffuse types of stomach-cancer – a multicenter case-control study in Poland. Cancer *Causes Control,* 2:227-233.

Bonaccorsi P, Caristi C, Gargiulli C, Leuzzi U (2008). Flavonol glucosides in *Allium* species: A comparative study by means of HPLC-DAD-ESI-MS-MS. *Food Chemistry,* 107:1668-1673.

Boots AW, Haenen GRMM, Bast A (2008). Health effects of quercetin: From antioxidant to nutraceutical. *European Journal of Pharmacology,* 585:325-337.

Boyle SP, Dobson VL, Duthie SJ, Kyle JAM, Collins AR (2000). Absorption and DNA protective effects of flavonoid glycosides from an onion meal. European. *Journal of Nutrition,* 39:213-223.

Briggs WH, Folts JD, Osman HE, Goldman IL (2001). Administration of raw onion inhibits platelet-mediated thrombosis in dogs. *Journal of Nutrition,* 131:2619-2622.

Caridi D, Trenerry VC, Rochfort S, Duong S, Laugher D, Jones R (2007). Profiling and quantifying quercetin glucosides in onion (*Allium cepa* L.) varieties using capillary zone electrophoresis and high performance liquid chromatography. *Food Chemistry,* 105:691-699.

Choi SI, Hong EY, Lee JH, Lee YS, Kim GH (2008). Antioxidant and antimicrobial activities of the ethanol extract of *Allium victoriallis* L. var. *platyphyllum*. *Food Science and Biotechnology,* 17:313-318.

Corea G, Fattorusso E, Lanzotti V, Capasso R, Izzo AA (2005). Antispasmodic saponins from bulbs of red onion, *Allium* cepa L. var. Tropea. *Journal of Agricultural and Food Chemistry,* 53:935-940.

Davis JM, Murphy EA, Carmichael MD, Davis B (2009). Quercetin increases brain and muscle mitochondrial biogenesis and exercise tolerance. *American Journal of Physiology-Regulatory Integrative and Comparative Physiology*, 296:R1071-R1077.

Day AJ, Mellon F, Barron D, Sarrazin G, Morgan MRA, Williamson G (2001). Human metabolism of dietary flavonoids: Identification of plasma metabolites of quercetin. *Free Radical Research, 35*:941-952.

Ding C-K, Chachin K, Ueda Y, Wang CY (2002). Inhibition of loquat enzymatic browning by sulfhydryl compounds. *Food Chemistry,* 76:213-218.

El-Demerdash FM, Yousef MI, Abou El-Naga NI (2005). Biochemical study on the hypoglycemic effects of onion and garlic in alloxan-induced diabetic rats. *Food and Chemical Toxicology,* 43:57-63.

Ernst M, Feldheim W (2000). Fructans in higher plants and in human nutrition. *Journal of Applied Botany and Food Quality–Angewandte Botanik,* 74:5-9.

Ewald C, Fjelkner-Modig S, Johansson K, Sjöholm I, Akesson B (1999). Effect of processing on major flavonoids in processed onions, green beans, and peas. *Food Chemistry,* 64:231-235.

FAO Statistics (2009). *Production. Crops.* Retrieved May 2011 from http://faostat.fao.org/site/567/default.aspx#ancor

Gabler NK, Osrowska E, Imsic M, Eagling DR, Jois M, Tatham BG, Dunshea FR (2006). Dietary onion intake as part of a typical high fat diet improves indices of cardiovascular health using the mixed sex pig model. *Plant Foods for Human Nutrition, 61*:179-185.

Galeone C, Pelucchi C, Levi F, Negri E, Franceschi S, Talamini R, Giacosa A, La Vecchia C (2006). Onion and garlic use and human cancer. *American Journal of Clinical Nutrition,* 84:1027-1032.

Galeone C, Tavani A, Pelucchi C, Negri E, La Vecchia C (2009). *Allium* vegetable intake and risk of acute myocardial infarction in Italy. *European Journal of Nutrition,* 48:120-123.

Gao CM, Takezaki T, Ding JH, Li MS, Tajima K (1999). Protective effect of *Allium* vegetables against both esophageal and stomach cancer: A simultaneous case-referent study of a high-epidemic area in Jiangsu province, China. *Japanese Journal of Cancer Research,* 90:614-621.

Glässer G, Graefe EU, Struck F, Veit M, Gebhardt R (2002). Comparison of antioxidative capacities and inhibitory effects on cholesterol biosynthesis of quercetin and potential metabolites. *Phytomedicine,* 9:33-40.

Goldman IL, Kopelberg M, Debaene JE, Schwartz BS (1996). Antiplatelet activity in onion (*Allium* cepa) is sulfur dependent. *Thrombosis and Haemostasis,* 76:450-452.

González CA, Pera G, Agudo A, Bueno-De-Mesquita HB, Ceroti M, Boeing H, Schulz M, Del Giudice G, Plebani M, Carneiro F, Berrino F, Sacerdotde C, Tumino R, Panico S, Berglund G, Simán H, Hallmans G, Stenling R, Martinez C, Dorronsoro M, Barricarte A, Navarro C, Quiros JR, Allen N, Key TJ, Bingham S, Day NE, Linseisen J, Nagel G, Overvad K, Jensen MK, Olsen A, Tjønneland A, Büchner FL, Peeters PH, Numans ME, Clavel-Chapelon F, Boutron-Ruault MC, Roukos D, Trichopolou A, Psaltopoulou T, Lund E, Casagrande C, Slimani N, Jenab M, Riboli E (2006). Fruit and vegetable intake and the risk of stomach and oesophagus adenocarcinoma in the European Prospective Investigation into Cancer and Nutrition (EPIC-EURGAST). *International Journal of Cancer,* 118:2559-2566.

González CA, Sanz JM, Marcos G, Pita S, Brullet E, Saigi E, Badia A, Riboli E (1991). Dietary factors and stomach-cancer in Spain - a multicenter case-control study. *International Journal of Cancer,* 49:513-519.

Gorinstein S, Leontowicz H, Leontowicz M, Jastrzebski Z, Najman K, Tashma Z, Katrich E, Heo B-G, Cho J-Y, Park Y-J, Trakhtenberg S (2010). The Influence of raw and processed garlic and onions on plasma classical and non-classical atherosclerosis indices: Investigations *in vitro* and *in vivo. Phytotherapy Research* 24:706-714.

Griffiths G, Trueman L, Crowther T, Thomas B, Smith B (2002). Onions-A global benefit to health. *Phytotherapy Research,* 16:603-615.

Guyonnet D, Belloir C, Suschetet M, Siess M-H, Le Bon A-M (2001). Antimutagenic activity of organosulfur compounds from *Allium* is associated with phase II enzyme induction. *Mutation Research: Genetic Toxicology and Environmental Mutagenesis*, 495:135-145.

Hansson LE, Nyrén O, Bergström R, Wolk A, Lindgren A, Baron J, Adami HO (1993). Diet and risk of gastric-cancer - a population-based case-control study in Sweden. *International Journal of Cancer,* 55:181-189.

Hollman PC, de Vries JH, van Leeuwen SD, Mengelers MJ, Katan MB (1995). Absorption of dietary quercetin glycosides and quercetin in healthy ileostomy volunteers. *American Journal of Clinical Nutrition,* 62:1276-1282.

Hu JF, La Vecchia C, Negri E, Chatenoud L, Bosetti C, Jia X, Liu R, Huang G, Bi D, Wang C (1999). Diet and brain cancer in adults: A case-control study in northeast China. *International Journal of Cancer,* 81:20-23.

Hubbard GP, Wolffram S, de Vos R, Bovy A, Gibbins JM, Lovegrove JA (2006). Ingestion of onion soup high in quercetin inhibits platelet aggregation and essential components of the collagen-stimulated platelet activation pathway in man: a pilot study. *British Journal of Nutrition,* 96:482-488.

Irkin R, Korukluoglu M (2007). Control of *Aspergillus niger* with garlic, onion and leek extracts. *African Journal of Biotechnology,* 6:384-387.

Iyengar R, McEvily AJ (1992). Anti-browning agents: alternatives to the use of sulfites in foods. *Trends in Food Science and Technology,* 3:60-64.

Jaime L, Martín-Cabrejas MA, Mollá E, López-Andréu FJ, Esteban RM (2001a). Effect of storage on fructan and fructooligosaccharide of onion (*Allium cepa* L.). *Journal of Agricultural and Food Chemistry,* 49:982-988.

Jaime L, Martínez F, Martín-Cabrejas MA, Mollá E, López-Andréu FJ, Waldron KW, Esteban RM (2001b). Study of total fructan and fructooligosaccharide content in different onion tissues. *Journal of the Science of Food and Agriculture,* 81:177-182.

Jaime L, Mollá E, Fernández A, Martín-Cabrejas MA, López-Andréu FJ, Esteban RM (2002). Structural carbohydrate differences and potential source of dietary fiber of onion (*Allium cepa* L.) tissues. *Journal of Agricultural and Food Chemistry,* 50:122- 128.

Jang MS, Sanada A, Ushio H, Tanaka M, Ohshima T (2002). Inhibitory effects of 'Enokitake' mushroom extracts on polyphenol oxidase and prevention of apple browning. *Lebensmittel-Wissenschaft und-Technologie,* 35:697-702.

Jung YS, Kim MH, Lee SH, Baik EJ, Park SW, Moon CH (2002). Antithrombotic effect of onion in streptozotocin-induced diabetic rat. *Prostaglandins, Leukotrienes and Essential Fatty Acids,* 66:453-458.

Kahane R, Vialle-Guérin E, Boukema I, Tzanoudakis D, Bellamy C, Chamaux C, Kik C (2001). Changes in non-structural carbohydrate composition during bulbing in sweet and high-solid onions in field experiments. *Environmental and Experimental Botany,* 45:73-83.

Kim JW, Huh JE, Kyung SH, Kyung KH (2004a). Antimicrobial activity of alk(en)yl sulfides found in essential oils of garlic and onion. *Food Science and Biotechnology,* 13:235-239.

Kim JW, Kim YS, Kyung KH (2004b). Inhibitory activity of essential oils of garlic and onion against bacteria and yeasts. *Journal of Food Protection,* 67:499-504.

Kim M-J, Kim CY, Park I (2005). Prevention of enzymatic browning of pear by onion extract. *Food Chemistry,* 89:181-184.

Kim SJ, Kim GH (2006). Quantification of quercetin in different parts of onion and its DPPH radical scavenging and antibacterial activity. *Food Science and Biotechnology,* 15:39-43.

Kumari K, Augusti KT (2007). Lipid lowering effect of S-methyl cysteine sulfoxide from *Allium cepa* Linn in high cholesterol diet fed rats. *Journal of Ethnopharmacology*, 109:367-371.

Kyung KH, Lee YC (2001). Antimicrobial activities of sulfur compounds derived from S-alk(en)yl-L-cysteine sulfoxides in *Allium* and *Brassica. Food Reviews International,* 17:183-198.

Langos M, Hofstetter W, Dolder S, Felix R, Mühlbauer RC, Brenneisen R (2007). A gamma-glutamyl peptide from onion inhibits the development and activity of osteoclasts *in vitro. Calcified Tissue International,* 80:S91-S92.

Lanzotti V (2006). The analysis of onion and garlic. *Journal of Chromatography A,* 1112:3-22.

Lara-Villoslada F, de Haro O, Camuesco D, Comalada M, Velasco J, Zarzuelo A, Xaus J, Galvez J (2006). Short-chain fructooligosaccharides, in spite of being fermented in the upper part of the large intestine, have anti-inflammatory activity in the TNBS model of colitis. *European Journal of Nutrition,* 45:418-425.

Lee M-K (2007). Inhibitory effect of banana polyphenol oxidase during ripening of banana by onion extract and Maillard reaction products. *Food Chemistry,* 102:146- 149.

Lee SK, Hwang JY, Kang MJ, Kim YM, Jung SH, Lee JH, Kim JI (2008a). Hypoglycemic effect of onion skin extract in animal models of diabetes mellitus. *Food Science and Biotechnology,* 17:130-134.

Lee SU, Lee JH, Choi SH, Lee JS, Ohnisi-Kameyama M, Kozukue N, Levin CE, Friedman M (2008b). Flavonoid content in fresh, home-processed, and light-exposed onions and in dehydrated commercial onion products. *Journal of Agricultural and Food Chemistry,* 56:8541-8548.

Ly TN, Hazama C, Shimoyamada M, Ando H, Kato K, Yamauchi R (2005). Antioxidative compounds from the outer scales of onion. *Journal of Agricultural and Food Chemistry,* 53:8183-8189.

Martínez MV, Whitaker JR (1995). The biochemistry and control of enzymatic browning. *Trends in Food Science and Technology,* 6:195-200.

Matheson EM, Mainous AG 3rd, Carnemolla MA (2009). The association between onion consumption and bone density in perimenopausal and postmenopausal non-Hispanic white women 50 years and older. *Menopause,* 16:756-759.

McAnlis GT, McEneny J, Pearce J, Young IS (1999). Absorption and antioxidant effects of quercetin from onions, in man. *European Journal of Clinical Nutrition,* 53:92-96.

Mogren LM, Caspersen S, Olsson ME, Gertsson U (2008). Organically fertilized onions (*Allium cepa* L.): Effects of the fertilizer placement method on quercetin content and soil nitrogen dynamics. *Journal of Agricultural and Food Chemistry,* 56:361-367.

Mogren LM, Olsson ME, Gertsson UE (2007). Quercetin content in stored onions (*Allium cepa* L.): Effects of storage conditions, cultivar, lifting time and nitrogen fertiliser level. *Journal of the Science of Food and Agriculture,* 87:1595-1602.

Moon JH, Nakata R, Oshima S, Inakuma T, Terao J (2000). Accumulation of quercetin conjugates in blood plasma after the short-term ingestion of onion by women. American Journal of Physiology-Regulatory, *Integrative, and Comparative Physiology,* 279:R461-R467.

Moreno FJ, Corzo-Martínez M, del Castillo MD, Villamiel M (2006). Changes in antioxidant activity of dehydrated onion and garlic during storage. *Food Research International,* 39:891-897.

Moriarty RM, Naithani R, Surve B (2007). Organosulfur compounds in cancer chemoprevention. *Mini Reviews in Medicinal Chemistry,* 7:827-838.

Mühlbauer RC, Lozano A, Reinli A (2002). Onion and a mixture of vegetables, salads, and herbs affect bone resorption in the rat by a mechanism independent of their base excess. *Journal of Bone and Mineral Research,* 17:1230-1236.

Mullen W, Edwards CA, Crozier A (2006). Absorption, excretion and metabolite profiling of methyl-, glucuronyl-, glucosyl- and sulpho-conjugates of quercetin in human plasma and urine after ingestion of onions. *British Journal of Nutrition,* 96:107-116.

Munday R, Munday CM (2001). Relative activities of organosulfur compounds derived from onions and garlic in increasing tissue activities of quinone reductase and glutathione transferase in rat tissues. *Nutrition and Cancer-An International Journal,* 40:205-210.

Munday R, Munday CM (2004). Induction of phase II enzymes by aliphatic sulfides derived from garlic and onions: An Overview. *Methods in Enzymology,* 382:449-456.

Negishi O, Ozawa T (2000). Inhibition of enzymatic browning and protection of sulfhydryl enzymes by thiol compounds. *Phytochemistry,* 54:481-487.

Nemeth K, Piskula MK (2007). Food content, processing, absorption and metabolism of onion flavonoids. *Critical Reviews in Food Science and Nutrition,* 47:397-409.

Norton T, Sun D-W (2008). Recent advances in the use of high-pressure as an effective processing technique in the food industry. *Food and Bioprocess Technology,* 1:2-34.

Nuutila AM, Kammiovirta K, Oskman-Caldentey K-M (2002). Comparison of methods for the hydrolysis of flavonoids and phenolic acids from onion and spinach for HPLC analysis. *Food Chemistry,* 76: 519-525.

Nuutila AM, Puupponen-Pimiä R, Aarni M, Oksman-Caldentey KM (2003). Comparison of antioxidant activities of onion and garlic extracts by inhibition of lipid peroxidation and radical scavenging activity. *Food Chemistry,* 81:485-493.

Ostrowska E, Gabler NK, Sterling SJ, Tatham BG, Jones RB, Eagling DR, Jois M, Dunshea FR (2004). Consumption of brown onions (*Allium cepa* var. *cavalier* and var. *density*) moderately modulates blood lipids, haematological and haemostatic variables in healthy pigs. *British Journal of Nutrition,* 91:211-218.

Paganga G, Miller N, Rice-Evans CA (1999). The polyphenolic content of fruit and vegetables and their antioxidant activities. What does a service constitute? *Free Radical Research,* 30:153-162.

Paltrinieri G, Figuerola F, Rojas L (1997). Technical manual on small-scale processing of fruits and vegetables. FAO. Retrieved May 2011 from: http://www.fao.org/docrep/X0209E/x0209e00.HTM

Park J, Kim J, Kim MK (2007). Onion flesh and onion peel enhance antioxidant status in aged rats. *Journal of Nutritional Science and Vitaminology,* 53:21-29.

Park SY, Yoo SS, Shim JH, Chin KB (2008). Physicochemical properties, and antioxidant and antimicrobial effects of garlic and onion powder in fresh pork belly and loin during refrigerated storage. *Journal of Food Science,* 73:C577-C584.

Powolny AA, Singh SV (2008). Multitargeted prevention and therapy of cancer by diallyl trisulfide and related *Allium* vegetable-derived organosulfur compounds. *Cancer Letters,* 269:305-314.

Roberfroid MB (2007). Inulin-type fructans: Functional food ingredients. *Journal of Nutrition,* 137:2493S-2502S.

Roldán E, Sánchez-Moreno C, de Ancos B, Cano MP (2008). Characterisation of onion (*Allium cepa* L.) by-products as food ingredients with antioxidant and antibrowning properties. *Food Chemistry,* 108:907-916.

Roldán-Marín E, Jensen RI, Krath BN, Kristensen M, Poulsen M, Cano MP, Sánchez-Moreno C, Dragsted LO (2010). An onion byproduct affects plasma lipids in healthy rats. *Journal of Agricultural and Food Chemistry,* 58:5308-5314.

Roldán-Marín E, Krath BN, Poulsen M, Binderup M-L, Nielsen TH, Hansen M, Barri T, Langkilde S, Cano MP, Sánchez-Moreno C, Dragsted LO (2009a). Effects of an onion by-product on bioactivity and safety markers in healthy rats. *British Journal of Nutrition,* 102:1574-1582.

Roldán-Marín E, Sánchez-Moreno C, Lloría R, de Ancos B, Cano MP (2009b). Onion high-pressure processing: Flavonol content and antioxidant activity. *LWT-Food Science and Technology,* 42:835-841.

Rose P, Whiteman M, Moore PK, Zhu YZ (2005). Bioactive S-alk(en)yl cysteine sulfoxide metabolites in the genus *Allium*: The chemistry of potential therapeutic agents. *Natural Product Reports,* 22:351-368.

Sakakibara H, Yoshino S, Kawai Y, Terao J (2008). Antidepressant-like effect of onion (*Allium cepa* L.) powder in a rat behavioral model of depression. *Bioscience, Biotechnology, and Biochemistry,* 72:94-100.

Sance MM, Gonzalez RE, Soto VC, Galmarini CR (2008). Relationships between antiplatelet activity, dry matter content and flavor in onion cultivars. *Journal of Food Agriculture and Environment,* 6:41-46.

Sankaranarayanan R, Varghese C, Duffy SW, Padmakumary G, Day NE, Nair MK (1994). A case-control study of diet and lung-cancer in Kerala, South-India. *International Journal of Cancer,* 58:644-649.

Santas J, Carbó R, Gordon MH, Almajano MP (2008). Comparison of the antioxidant activity of two Spanish onion varieties. *Food Chemistry,* 107:1210-1216.

Schieber A, Stintzing FC, Carle R (2001). By-products of plant food processing as a source of functional compounds-Recent developments. *Trends in Food Science and Technology,* 12:401-413.

Slimestad R, Fossen T, Vågen IM (2007). Onions: A source of unique dietary flavonoids. *Journal of Agricultural and Food Chemistry,* 55:10067-10080.

Slobodianik N, Insani M, Feliu MS, Galmarini CR (2007). Antioxidant effects from onion (*Allium cepa* L.) on oxidative stress of rats cardiac muscle. *FASEB Journal,* 21:A726.

Srinivasan K (2005). Plant foods in the management of diabetes mellitus: Spices as beneficial antidiabetic food adjuncts. *International Journal of Food Sciences and Nutrition,* 56:399-414.

Stajner D, Milić N, Canadanović-Brunet J, Kapor A, Stajner M, Popović BM (2006). Exploring *Allium* species as a source of potential medicinal agents. *Phytotherapy Research,* 20:581-584.

Takahama U, Hirota S (2000). Deglucosidation of quercetin glucosides to the aglycone and formation of antifungal agents by peroxidase dependent oxidation of quercetin on browning of onion scales. *Plant and Cell Physiolology,* 41:1021-1029.

Teyssier C, Amiot MJ, Mondy N, Auger J, Kahane R, Siess MH (2001). Effect of onion consumption by rats on hepatic drug-metabolizing enzymes. *Food and Chemical Toxicology,* 39:981-987.

Tuyns AJ, Kaaks R, Haelterman M, Riboli E (1992). Diet and gastric-cancer. A case-control study in Belgium. *International Journal of Cancer,* 51:1-6.

USDA National Nutrient Database for Standard Reference, Release 21 (2008). Retrieved May 2011 from: http://www.nal.usda.gov/fnic/foodcomp/Data/; http://www.nal.usda.gov/fnic/foodcomp/Data/SR21/reports/sr21fg11.pdf.

Vázquez-Prieto MA, Miatello RM (2010). Organosulfur compounds and cardiovascular disease. *Molecular Aspects of Medicine* 31:540-545.

Waldron, K (1999). Finding useful ingredients in onion waste. *Emerging Food* R&D Report.

Waldron, K (2001). Useful ingredients from onion waste. *Food Science and Technology,* 15:38-41.

Wang HX, Ng TB (2004). Isolation of allicepin, a novel antifungal peptide from onion (*Allium cepa*) bulbs. *Journal of Peptide Science,* 10:173-177.

Wetli HA, Brenneisen R, Tschudi I, Langos M, Bigler P, Sprang T, Schürch S, Mühlbauer RC (2005). A gamma-glutamyl-peptide isolated from onion (*Allium cepa* L.) by bioassay-guided fractionation inhibits resorption activity of osteoclasts. *Journal of Agricultural and Food Chemistry,* 53:3408-3414.

Woo KS, Hwang IG, Kim TM, Kim DJ, Hong JT, Jeong HS (2007).Changes in the antioxidant activity of onion (*Allium cepa*) extracts with heat treatment. *Food Science and Biotechnology,* 16:828-831.

Xiao H, Parkin KL (2007). Isolation and identification of potential cancer chemopreventive agents from methanolic extracts of green onion (*Allium cepa*). *Phytochemistry,* 68:1059-1067.

Yang J, Meyers KJ, Van der Heide J, Liu RH (2004). Varietal differences in phenolic content and antioxidant and antiproliferative activities of onions. *Journal of Agricultural and Food Chemistry,* 52:6787-6793.

You WC, Blot WJ, Chang YS, Ershow A, Yang ZT, An Q, Henderson BE, Fraumeni JF Jr, Wang TG (1989). *Allium* vegetables and reduced risk of stomach-cancer. *Journal of the National Cancer Institute,* 81:162-164.

Yuan L, Ji TF, Wang AG, Yang JB, Su YL (2008). Two new furostanol saponins from the seeds of *Allium cepa* L. *Chinese Chemical Letters,* 19:461-464.

In: Onion Consumption and Health
Editors: C. B. Aguirre et al.
ISBN: 978-1-62100-836-1

Chapter 5

HEALTHY PROPERTIES FROM ONION PRODUCTS

Vanesa Benítez, Esperanza Mollá, María A. Martín-Cabrejas, Yolanda Aguilera, Rosa M. Esteban
Instituto de Investigación en Ciencias de la Alimentación (CIAL) CSIC-UAM/Departamento de Química Agrícola, Facultad de Ciencias, Campus de la Universidad Autónoma de Madrid, Madrid, Spain

ABSTRACT

Onion (*Allium* cepa L.) is a food ingredient widely used in worldwide gastronomy, making a significant nutritional contribution to the human diet. Moreover, since ancient times, onions have been used as medicinal agents. In recent years, extensive research has focused on the beneficial and medicinal properties of onions and it has been reported that onions are effective in cardiovascular disease, due to their hypocholesterolemic, hypolipidemic, anti-hypertensive, anti-diabetic, antithrombotic and anti-hyperhomocysteinemia effects, and have many other biological activities such as antimicrobial, antioxidant, anticarcinogenic, antimutagenic, antiasthmatic, immunomodulatory and prebiotic activities. These properties are due to the presence in onions of several groups of compounds such as flavonoids, alk(en)yl cysteine sulphoxides, fructooligosaccharides and dietary fiber among others. Thus, onions and their derived products, such as onion skins, two outer fleshy scales and roots, constitute an interesting source of dietary fiber, phytochemicals and natural antioxidants, so that their application in food, which would increase health promoting properties of foods, is a promising field. However, the composition varies throughout the bulb and also is dependent of the cultivar. Therefore, in the present article, healthy properties of products derived from onions have been studied and reviewed in relation to their composition, in order to assess their possible use as functional ingredients rich in specific compounds. Such information may be useful to food technologists for the appropriate exploitation of onion products as source of a specific functional compound.

INTRODUCTION

Onion (*Allium cepa* L.), botanically included in the Liliaceae family, is a species of great economic importance, widely cultivated all over the world due to its great diversity which make them able to adapt to a very wide range of environments. Onion is thought to have originated in central Asia. Since ancient times onions have been used as common foods and for the treatment of many diseases. The first citation of this plant is found in the Codex Ebers (1550 bc), an Egyptian medical papyrus reporting several therapeutic formulas based on onions as useful remedy for a variety of diseases such as heart problems, headache, bites, worms and tumours. Later on, this food plant was known by Greeks and Romans, that used it as important healing agents just as it still is used from most of the people of the Mediterranean area (Lanzotti, 2006).

Onions are the second most important horticultural crop worldwide, after tomatoes, with current annual production around 66 million tonnes. Over the past 10 years, onion production has increased by more than 25% (FAO, 2009). Moreover, onion consumption is increasing significantly due to its significant nutritional contribution to the human diet, and also to its medicinal and functional properties. Onions are versatile and are used as an ingredient in many dishes and accepted by almost all traditions and cultures. In Europe, onion is one of the main vegetables consumed either raw or processed in different ways (Benítez et al., 2011).

Onion contains several groups of substances that have been suggested to have health promoting effects and as it is commonly consumed, it could be a valuable source of health promoting compounds. Onion extracts have been recently reported to be effective in cardiovascular disease, because of their hypocholesterolemic, hypolipidemic, anti-hypertensive, anti-diabetic, antithrombotic and anti-hyperhomocysteinemia effects, and to possess many other biological activities including antimicrobial, antioxidant, anticarcinogenic, antimutagenic, antiasthmatic, immunomodulatory and prebiotic activities (Corzo-Martínez et al., 2007). For example, onion has been found to be associated with a lower risk for lung cancer and for esophageal and stomach cancer (Gao et al., 1999; Le Marchand, 2002). Moreover, in an experiment using pigs as a biomedical model, onion was found to reduce total blood cholesterol, low density lipoprotein, and triglycerides, which are used as indicators of cardiovascular disease risk (Gabler et al., 2005). Furthermore, onion showed antithrombotic activity both *in vitro* and *in vivo* in mice (Yamada et al., 2004) and an antithrombotic effect in streptozotocin-induced diabetic rats (Jung et al., 2002; Olsson et al., 2010).

Lately, there has been an increase in demand for processed onions which has led to an increase in waste production. Accordingly more than 500,000 tonnes of onion waste are produced annually in the European Union, mainly from Spain, UK and Holland. These residues come mainly from the market demand of peeled and processed onions such as onion ring, frozen cut onion and freeze dried onion for ready to eat food. The major industrial waste are the outer two fleshy leaves and lesser quantities of onion brown skin and tops and bottoms of bulbs. Moreover, onion waste also include undersized, malformed, diseased or damaged bulbs. Nowadays, worthless vegetables represented a relevant problem because food industries and farmers are forced by environmental regulations to develop productions without secondary residues. Therefore, there is a considerable emphasis on the recovery, recycling and upgrading of waste. They are usually utilized to feed animals, and also they can be used as organic fertilizer. However, onion waste is not suitable to be used as fodder due to their strong aroma, and neither they can be used as organic fertilizers because of their susceptibility to phytopathogens such as *Sclerotium cepivorum* (white rot) (Schieber et al., 2001). Moreover, their removal by combustion becomes rather expensive as a result of their high percentage of moisture. Thus, onion producers and processors are all interested in developing alternative means for the valorization of the onion waste to promote its profitable usage and its subsequent conversion into food-grade products due to well-known health properties of onions (González-Sáiz et al., 2008). The waste has been identified as a potential source of flavour compounds, dietary fibre (DF) components, non-structural carbohydrates like fructans and fructooligosaccharides, and flavonoids particularly quercetin glycosides (Jaime et al., 2001a; Jaime, et al., 2002; Rose et al., 2005; Roldán et al, 2008; Benítez et al., 2011). However, onion composition is variable and depends on cultivar, stage of maturation, environment, agronomic conditions, storage time and bulb section. Therefore, it is necessary to know the composition of each industrial onion waste to know its potential health benefits.

In general, water makes up the majority (80–95%) of the fresh weight of onion. Up to 65% or more of the dry weight may be in the form of non-structural carbohydrates (NSC) which include glucose, fructose, sucrose and FOS (Davis et al., 2007). The main FOS in onion bulbs are kestose, nystose and fructofuranosylnystose (Jaime et al., 2001). Onion also shows an important quantity of total dietary fiber (TDF) and a good soluble: insoluble dietary fiber ratio (SDF: IDF). Moreover, onion is known for its flavonoid content, contributing considerably to its dietary intake in many countries (Santas et al., 2008). Two flavonoid subgroups are present in onions; anthocyanins, which impart a red/purple colour to some varieties and flavonols such as quercetin and its derivatives which may play a role in the production of yellow and brown compounds of many other varieties (Downes et al., 2009). In recent literature, quercetin 4'-glucoside and quercetin 3,4'-diglucoside are in most cases reported as the main onion flavonols of the flesh (Roldán-Marín et al., 2009; Downes et al., 2010), whereas onion skins contain higher concentrations of quercetin aglycon (Downes et al., 2009). Likewise, alk(en)yl cisteine sulphoxides (ACSOs) are the flavour and aroma precursors which, when cleaved by the enzyme alliinase, generate the characteristic odour and taste of onion. Four ACSOs have been identified in *Alliums*, and the flavour variation among species is due to differences in ACSO composition and concentration (Randle et al., 1995). The three naturally occurring ACSOs in onion are trans-(+)-S-1-propenyl-L-cysteine sulphoxide (PECSO), which is normally found in the highest concentration and gives rise to the compound responsible for the lachrymatory effect, and (+)-S-methyl-L-cysteine sulphoxide (MCSO) and (+)-S-propyl-L-cysteine sulphoxide (PCSO), which are found in smaller amounts (Mallor and Thomas, 2008).

In the present article, healthy properties of products derived from onions, brown skin, top and bottom, outer two fleshy scales, inner scales and whole onion, have been studied and reviewed in relation to their composition in order to assess their possible use as functional ingredients rich in specific compounds.

BIOACTIVE CHARACTERIZATION OF PRODUCTS DERIVED FROM ONION

Dietary Fiber

DF content of onion is cultivar dependent, thus a wide range of DF is found among onion cultivars (Table 1). For example, cultivars such as Grano de Oro, Hysam or Recas have higher DF content than Sturon or Figueres (Jaime et al., 2002; Benítez et al., 2011). Brown skin shows the highest content of TDF, followed by top-bottom which might suggest a decrease of TDF from the outer to the inner of the bulb. Top-bottom product, obtained by slicing off 5–10 mm the top and bottom ends of the bulb, is formed by a mixture of inner, outer and skin sections; thus the higher amounts of TDF could be due to the high contribution of skin to the top. It is interesting to point out the different values found in outer and inner scales. Apparently, these products seem to show the same physical characteristics; however, TDF values were higher in outer fleshy scales than in inner ones. This tendency may suggest higher cell wall development in outer scales (Jaime et al., 2002; Benítez et al., 2011).

IDF is the main fraction of TDF in all onion products and, therefore, it showed the same trend found for TDF. The main polysaccharides of onion IDF are cellulose and polyuronides, although there are differences between products. Thus, IDF from outer two fleshy scales comprise mainly pectic polysaccharides, whereas the inner scales IDF is mainly formed by cellulose. On the other hand, the carbohydrate composition of top- bottom show the presence of pectic polysaccharides, as inferred from the percentages of uronic acid, galactose, and arabinose, and the presence of cellulose. Moreover, the levels of xylose and arabinose in top-bottom products are higher than those found in other onion products. The sugar profile of the skin IDF exhibit lower levels of galactose when compared to other onion product. The lower galactose content might increase the cross-linking between pectic polysaccharides and, hence, the packing of pectin strands, which has been related to the firmness development of cells and waterproofing characteristics of the onion skin (Jaime et al., 2002).

SDF contents are significantly lower than IDF contents in all onion products, being top-bottom the product which shows the highest SDF content. The carbohydrate composition of SDF fraction shows the presence of uronic acid and galactose as the main sugar constituents, whereas arabinose appears in minor amounts, and only traces of xylose were detected in these products.

Brown skin shows lowest levels of galactose as in the IDF fraction. The galactose percentage decrease from inner to outer tissues in both fiber fraction in contrast to uronic acids behaviour. These trends indicate that higher amounts of rhamnogalacturonans substitute with galactans are found in fleshy scales (Jaime et al., 2002).

Table 1. Bioactive composition of onion products

mg g^{-1} DM	Whole onion	Inner scales	Outer scales	Top-bottom	Brown skin
IDF*	92-266	57-179	99-270	322-579	593-698
SDF*	41-67	35-52	42-59	54-90	17-82
TDF*	149-310	93-222	145-312	387-667	637-750
Kestose**	11-45	13-75	8-60	4-19	0.8
Total FOS**	28-73	27-139	16-100	6-29	0.8
Total Fructans**	42-273	45-316	23-281	8-56	22.4
NSC**	379-509	537-625	287-469	120-211	130.3
Total Phenolics***	17.3	9.4	19.7	30.5	52.7
Total Flavonoids***	10.3	7.0	19.5	25.9	43.1
Total Flavonols***	9.0	6.1	19.2	15.3	7.9
Quercetin 4'-glucoside***	4.0	2.0	7.4	6.3	5.2
Quercetin 3,4'-diglucoside***	3.1	3.7	9.5	5.9	0.3
Quercetin***	0.9	0.0	0.6	1.2	1.6
Total ACSOs****	4.2	8.1-11.4	5-15.7	3.7-13.1	0.7-2.2
PECSO****	2.2	6.0-10.6	3.6-15	2.4-12.5	0.4-2.2
MCSO****	2.0	0.8-3.1	0.6-1.4	0.4-1.3	0.0-0.3

* Jaime et al., 2002; Benítez et al., 2011.

** Jaime et al., 2001b; Benítez et al., 2011.

*** Benítez et al., 2011.

**** Bacon et al., 1999; Benítez et al., 2011.

Soluble/insoluble fiber ratios are important from both dietary and functional perspectives. To be acceptable, a dietary fiber ingredient must perform in a satisfactory manner as a food ingredient. It must be kept in mind that fiber enrichment not only influences the overall quality of food by changing its physiological properties but also significantly affects the sensorial properties of a product. Besides the amount of DF added, the ratio of insoluble to soluble fiber is an important variant related to structural and also sensorial properties (Esteban et al., 1998). It is generally accepted that those fiber sources suitable to be used as a food ingredient should supply a balanced amount of approximately 30% of soluble dietary fiber (Grigelmo-Miguel and Martín-Belloso, 1999). In this respect, inner scales and outer fleshy scales provide the more suitable onion products for food supplementation, whereas brown skin and top-bottom show high TDF contents which is very interesting for their potential industrial use as a fiber source, although it would be necessary some treatment in order to solubilise part of IDF fraction (Waldron, 2001).

Consumption of dietary fiber (DF) has been reported to be effective in lowering the risk of cardiovascular disease, gastrointestinal diseases, colon cancer, type 2 diabetes and obesity (Champ et al., 2003). These beneficial effects might be globally explained by the mechanisms of action of DF. Thus, DF consumption produce a reduction of energy intake, a satiating effect due to its viscosity and gel formation, as well as a decreased absorption of other food constituents (carbohydrates, protein, fats) in the upper gastrointestinal tract, due to the increase of intestinal bulk and the decrease of transit time which also make difficult the access of digestive enzymes to its substrates. Recent findings have demonstrated that dietary fiber derived from some fruits and vegetables may promote a significant decrease in blood cholesterol concentration (Chau et al., 2004). The reductions in the levels of cholesterol and lipids by DF consume can be a consequence of reduced cholesterol biosynthesis, change of bile acid synthesis, and reduced absorption of lipid, cholesterol and bile acids (Chau et al., 2004).

Non-structural Carbohydrates

Non-structural carbohydrates consist in fructans, sucrose, glucose and fructose (Table 1). The study of the non-structural carbohydrate profile of bulbs from different onion cultivars revealed that fructans are the main carbohydrate fraction in cultivars such as Sturon, Hysam, Durco and Caribo, while cultivars such as Grano de Oro, Figueres and Recas contain higher levels of free glucose and fructose (Jaime et al., 2001a; Benítez et al., 2011). Therefore, fructan content depends on the cultivar. Cultivars with high dry matter are found to contain high levels of fructans and vice versa. It has been suggested that the different distribution patterns among cultivars are genetically controlled. According to Jaime et al. (2001a) the profile of fructan distribution in onion bulbs suggests that these compounds have been hydrolyzed to fructose in low dry weight varieties to facilitate osmo-regulation as the bulb takes up water and expands during bulb developing, and water content is therefore higher than in high dry weight cultivars where fructans are not hydrolyzed because of their genetically controlled inability to take up water. Regarding FOS content and composition, the total content of FOS as sum of kestose, nystose and 1-F-nystose varied considerably depending on the onion cultivar. However, in general, the prevalence of the lowest polymerized FOS (kestose) was observed.

Regard to onion products, the highest concentration of NSC, total fructans and FOS are found in the inner scales, followed by outer two fleshy leaves. However, it has been found differences between inner scales and outer two fleshy scales with regard to NSC content and contributions, since sucrose and fructans contribute less to NSC content in outer scales than in inner. The distribution of FOS in the bulbs suggests that these compounds have been hydrolyzed to fructose in the outer scales, with lower dry matter, to stimulate osmotic water uptake as the onion expands during bulb development. The water content is therefore higher in outer than in the inner scales where the extension of hydrolysis is lower (Slimestad and Vagen, 2009). Top-bottom product and especially brown skin show the lowest levels of FOS (Jaime et al., 2001b). Moreover, the NSC profile in top-bottom is different, since fructose constituted the main sugar and sucrose contribution was higher than in fleshy wastes.

The content of FOS decreases as the degree of polymerisation increase, being kestose the main FOS in all onion products. Onion fructans are composed mainly by FOS of low degree of polymerization (DP3-DP5). In onion products, FOS contribution to total fructans is more than 60% and it is observed that the higher is the fructan content, the greater is its degree of polymerisation (Benítez et al., 2011).

The use of FOS as a food ingredient has triggered much research on their possible health effects; in these sense, FOS are included in the prebiotics classification (Gibson and Roberfroid, 1995). FOS resist hydrolysis by digestive enzymes and enter the caecum and colon intact. No change in blood glucose is noted after oral ingestion, and they are considered safe for diabetics. Moreover, FOS have a low sweetness intensity, since they are only about one third as sweet as sucrose. This property is quite useful in various kinds of foods where the use of sucrose is restricted by its high sweetness. When FOS reach the large intestine, they are highly fermented, mainly by Bifidobacteria, into short-chain carboxylic acids (acetate, propionate, butyrate), which improve the intestinal flora avoiding colonization by pathogens. This property, together with their effect on gastric emptying and intestinal motility, allows them to be classified as soluble dietary fibres. FOS offer several important physiological properties, such as decreasing the levels of serum cholesterol, phospholipids and triglycerides, and they are also non-cariogenic substances (Jaime et al., 2001b).

Phenolic Compounds

The range of phenolic compounds is wide, since the contents of phenolic substances vary with the cultivars and also are influenced by a large number of external factors such as agrotechnical processes, climatic conditions and ripeness during harvest and postharvest manipulations (Stratil et al., 2006).

Regard to onion products (Table 1), brown skin shows the highest levels of total phenolics and flavonoids, and inner scales the lowest. Therefore, a decrease of total phenolics and flavonoids is observed from the outer to the inner part of the bulb (Patil and Pike, 1995; Prakash et al., 2007; Beesk et al., 2010; Benítez et al., 2011). This tendency suggests an increase of these compounds during aging of cells that constitute scales of onion bulbs because it is known that cells of the outer scales are more aged than those of the inner scales in a bulb and that cells of the upper position in a scale (Hirota et al, 1998). Moreover, flavonoids are the major group of phenolic compounds in onion and onion products (Yang et al., 2004; Santas et al., 2008; Benítez et al., 2011).

In relation to flavonols, the outer two fleshy scales show the highest level of flavonols, while brown skin and inner scales show the lowest amount of these compounds. A possible explanation for this distribution is the light-induced enzyme phenylalanine ammonia-lyase, which catalyses the biosynthesis of flavonoids. The outermost localising living cells of the whole onion bulb are in the middle scales, and in contrast to the dried and dead cells of the brown skin are able to perform the active biosynthesis of flavonoids. In contrast to the inner layers, the cells of outer scales are more influenced by light, therefore the light-induced enzyme phenylalanine ammonia-lyase could catalyse the flavonoid biosynthesis in the middle layers and contribute to higher flavonoid levels (Beesk et al., 2010). Trammell and Peterson (1976) studied the distribution of flavonols in the bulb and they found that the flavonol content in general increase from bottom to top (vertical distribution) of the bulbs by less than a factor of 2. The horizontal distribution shows a 2- to 3-fold increase in concentration from the centre of the bulb outward. Flavonols are the main components of flavonoids in fleshy scales, however, in top-bottom and, especially, in brown skin flavonols represent a small percentage of total flavonoids (Benítez et al., 2011). Flavonols can protect plant cells from UV light, since about 90% of total flavonols has been confined to epidermal tissue. Parenchymous storage tissue, which makes up the bulk of a bulb, contains only about 10% of the total pigment. It follows that scale thickness or any other factors which alter the ratio of epidermal to storage tissue could indirectly affect gross flavonol concentration (Hirota et al., 1998; Slimestad and Vagen, 2009).

The main flavonols are quercetin 4'-glucoside (QMG) and quercetin 3,4'-diglucoside (QDG) and the minor conjugates are isorhamnetin glucosides and quercetin 3-glucoside. Main flavonols account for over 80% of total flavonols in whole onion, inner and outer two fleshy scales and top-bottom. The diglucoside:monoglucoside ratio differ between onion products; a decrease of the ratio is observed from the inner to the outer of the bulb, which suggests that during aging of scales, monoglucoside is synthesized more rapidly than diglucoside or that glucose at the C-3 position of diglucoside is enzymatically hydrolyzed, producing monoglucoside as discussed by Tsushida and Suzuki (1996). The ratio between QDG and QMG in different onion products might be an important aspect from the viewpoint of food processing. Rohn et al. (2007) demonstrated that a profile high in QDG concentration seems to be advantageous under drying and roasting conditions, as the monoglucosidic intermediate QMG is formed having a high bioavailability (Beesk et al., 2010).

Free quercetin is found mainly in brown skin and top-bottom (Hirota et al., 1998; Slimestad and Vagen, 2009; Beesk et al., 2010; Benítez et al., 2011). The high quercetin content in brown skin could be due to the hydrolysis of quercetin glucosides during peel formation. This transformation suggests that quercetin is included in the formation of brown compounds in the dry skin (Patil and Pike, 1995; Hirota et al., 1998).

The presence of flavonoids in onion products confers them some healthy properties. For example, flavonoids are shown to have antioxidative activity, free radical scavenging capacity, coronary heart disease prevention, and anticancer activity. Moreover, some flavonoids exhibit potential for anti–human immunodeficiency virus functions. Furthermore, quercetin is known for its anticancer, antiinflamatory and antiviral activity (Formica and Regelson, 1995; Yao et al., 2004).

S-Alk(en)yl-L-cysteine sulphoxides

The highest S level is found in the inner scales and the lowest one in brown skin. Sulphur is incorporated into onion flavour precursors (ACSOs) among other compounds. However, there is no correlation between total S content and flavour precursor content (Benítez et al., 2011). Therefore, sulphur accumulation is poorly correlated with pungency in onion cultivars (Chope and Terry, 2009). Matikkala and Virtaten (1967) found that the total content of flavor precursor in onion ranged from 35 to 385 mg 100 g^{-1}FW.

In the studies which research ACSOs in onion products only two ACSOs were detected (Table 1), the (+)-S-methyl-l-cysteine sulphoxide (MCSO) and trans-(+)-S-(1-propenyl)-l-cysteine sulphoxide (PECSO). Propyl cysteine sulphoxide (PCSO) was not found in those cultivars (Recas, Grano de Oro, Hysam and Durco), since occasionally PCSO has not been detected in onion (Matikkala and Virtaten,1967; Thomas and Parkin, 1994). ACSO content shows good correlation with fructans and dry weight. Thus, generally low dry weight onions have low ACSO content (Chope et al., 2006).

On a dry weight basis, the inner scales contain the highest content of ACSOs, both PECSO and MCSO, followed by outer two fleshy scales and top-bottom. The lowest levels of precursors occurs in brown skin suggesting that this material is of limited value as a source of flavour compounds. It is not clear whether the low level of flavour precursors observed in the brown skin is due to contamination of this fraction with small amounts of tissue of intermediate scales consisting of both brown and white material (Bacon et al., 1999). Therefore, a decreasing gradient of concentration of flavour precursor from inner of the bulb to outer exist (Benítez et al., 2011). Randle (1997) indicates that flavour is unequally distributed within the onion plant and there is a flavour gradient within the bulb. Thus, the highest concentration of precursors occurs in the interior base of the bulb, while the lowest concentration of precursors is in the top outer scales. However, other authors (Bacon et al. 1999) found in outer two fleshy scales the highest content (on a dry weight basis) of flavour precursors.

PECSO, a precursor of the lachrymatory factor, is the main flavour precursor in onion products. The PECSO content depends on the variety and this content is related to the pungency, thus within onion species, there is a great difference in PECSO content between mild cultivars (0.78 mg g^{-1} FW) and extremely pungent cultivars (2.32 mg g^{-1} FW) (Yoo and Pike, 1998). Flavour precursor composition and concentration can significantly influence flavour and flavour intensity. The different precursor ratios give rise to different taste and aroma. For example, if total precursor concentrations were equal, a cultivar with a ratio of 12:4:2 (PECSO:MCSO:PCSO) would be more pungent and tear producing than another cultivar with 6:9:3 ratio (Randle, 1997). When the bulb is cut the enzyme alliinase converts ASCOs into different compounds such as pyruvate, 1-propenylsulfenic acid and ammonia. Some authors (Abayomi and Terry, 2009) measured the spatial distribution of pyruvate in cv SS1 and they found the same trend found by Benítez et al. (2011) for total S, MCSO, PECSO and total ACSO content suggesting pyruvate concentrations could be directly related to ACSO content.

Health benefits of *Allium* organosulphur compounds have been recently reviewed (Rose et al., 2005). They inhibit the aggregation of human blood platelets and offer the potential for positive cardiovascular health benefits. Furthermore, these compounds have the ability to positively modify the antioxidant, apoptotic and inflammatory systems in mammal (Rose et al., 2005).

Conclusion

The results show that each waste would have a profit. Thus, brown skin and top-bottom could be potentially used as functional ingredient rich in dietary fiber, mainly the insoluble fraction, and in total phenolics and flavonoides. Moreover, brown skin shows a high concentration of quercetin aglycone. On the other hand, outer two fleshy scales could be used as source of flavonols and dietary fiber content, with a good ratio SDF:IDF. However, inner scales could be an interesting source of fructans and ACSOs. In addition, discarded onions could be used as good source of dietary fiber and fructans. It would be interesting that processors will separate the different parts of onion obtained during the onion processing due to their added value. However, up to date they have not been always effectively exploited. Industrial processing of onions to obtain onion products, such as frozen cut onions and freeze dried onions for ready to eat food, implies the separation of the top and bottom of the onion bulbs and also the two outer fleshy scales together with the brown skin. Top and bottom constitutes a separated waste that could be potentially used as functional ingredient. Two outer fleshy scales and brown skin could be used together as a source of phenolics with good antioxidant activity. Nevertheless, the different composition of these two products would suggest the great interest of separating both products to exploit them as source of different bioactive compounds. Onion products, as it has been mentioned, constitute an interesting source of phytochemicals and natural antioxidants and their application in food, which increases health promoting properties of foods, is a promising field. Moreover, if the recovery and the production of new products from onion products are successful, the environmental problems could be solved. Furthermore, future investigations on the bioactivity, bioavailability and toxicology of onion product phytochemicals and their stability and interactions with other food ingredients should be carried out and carefully assessed by *in vitro* and *in vivo* studies. Undoubtedly, functional foods represent an important, innovative and rapidly growing part of the overall food market.

References

Abayomi, L.A. & Terry L.A. (2009). Implications of spatial and temporal changes in concentration of pyruvate and glucose in onion (Allium cepa L.) bulbs during controlled atmosphere storage. *J Sci Food Agric* 89: 683-687.

Bacon, J.R.; Moates, G.K.; Ng, A.; Rhodes, M.J.C.; Smith, A.C.; Waldron, K.W. (1999). Quantitative analysis of flavour precursors and pyruvate levels in different tissues and cultivars of onion (Allium cepa). *Food Chem* 64: 257-261.

Beesk, N.; Perner, H.; Schwarz, D.; George, E.; Kroh, L.W.; Rohn, S. (2010). Distribution of quercetin-3,4′-O-diglucoside, quercetin-4′-O-monoglucoside, and quercetin in different parts of the onion bulb (Allium cepa L.) influenced by genotype. *Food Chem* 122: 566-571.

Benítez, V.; Mollá, E.; Martín-Cabrejas, M.A.; Aguilera, Y.; López-Andréu, F.J.; Cools, K.; Terry, L.A.; Esteban R.M. (2011). Characterization of Industrial Onion Wastes (Allium cepa L.): Dietary Fiber and Bioactive Compounds. *Plant Food Hum Nutr* 66: 48–57.

Champ, M.; Langkilde, A.M.; Brouns, F.; Kettlitz, B.; Collet, Y.L.B. (2003) Advances in dietary fibre characterisation. 1. Definition of dietary fibre, physiological relevance, health benefits and analytical aspects. *Nutr Res Rev 16*: 71-82.

Chau, C.F.; Huang Y.L.; Lin C.Y. (2004). Investigation of the cholesterol-lowering action of insoluble fiber derived from the peel of Citrus sinensis L. cv. Liucheng, *Food Chem* 87: 361–366.

Chope, G.A. & Terry, L.A. (2009). Use of canonical variate analysis to differentiate onion cultivars by mineral content as measured by ICP-AES. *Food Chem* 115: 108-1113.

Chope, G.A.; Terry, L.A.; White, P.J. (2006). Effect of controlled atmosphere storage on abscisic acid concentration and other biochemical attributes of onion bulbs. *Postharvest Biol Technol* 39: 233-242.

Corzo-Martínez, M.; Corzo, N.; Villamiel, M. (2007). Biological properties of onions and garlic. *Trends Food Sci Tech* 18: 609-625.

Davis, F.; Terry, L.A.; Chope, G.A.; Faul, C.F.J. (2007). Effect of Extraction Procedure on Measured Sugar Concentrations in Onion (Allium cepa L.) Bulbs. *J Agric Food Chem* 55: 4299-4306.

Downes, K.; Chope G.A.; Terry, L.A. (2010). Postharvest application of ethylene and 1-methylcyclopropene either before or after curing affects onion (Allium cepa L.) bulb quality during long term cold storage. *Postharvest Biol Technol* 55: 36-44.

Downes, K.; Chope, G.A.; Terry, L.A. (2009). Effect of curing at different temperatures on biochemical composition of onion (Allium cepa L.) skin from three freshly cured and cold stored UK-grown onion cultivars. *Postharvest Biol Technol* 54: 80–86.

Esteban, R.M.; Mollá, E.; Valiente, C.; Jaime, L.; Lopez-Andréu, F.J.; Martin-Cabrejas, M.A. (1998). Dietary fibre: chemical and physiological aspects. *Rec Res Dev Agric & Food Chem* 2: 293-308.

FAO Statistics (2009) Productions, crops. Retrieved February 15 from http://faostat.fao.org/site/567/default.aspx#ancor.

Formica, J.V. & Regelson, W. (1995) Review of the Biology of Quercetin and Related Bioflavonoids. *Food Chem Toxicol* 33: 1061-1080.

Gabler, N.K.; Ostrowska, E.; Sterling, S. J.; Jones, R. B.; Tatham, B.G.; Eagling, D. R.; Jois, M.; Dunshea, F. R. (2005). Consumption of raw brown onions variably modulate plasma lipid profile and lipoprotein oxidation in pigs fed a high-fat diet. *J Sci Food Agric* 85: 154–160.

Gao, C.M.; Takezaki, T.; Ding, J.H.; Li, M.S.; Tajima, K. (1999). Protective Effect of Allium Vegetables against Both Esophageal and Stomach Cancer: A Simultaneous Case-referent Study of a High-epidemic Area in Jiangsu Province, China. *Cancer Sci* 90: 614–621.

Gibson, G.R. & Robertfroid, M.B. (1995). Dietary Modulation of the Human Colonic Microbiota: Introducing the Concept of Prebiotics. *J Nutr* 125: 1401-1412. .

González-Sáiz, J.M.; Esteban-Díez, I.; Rodríguez-Tecedor, S.; Pizarro, C. (2008). Valorization of Onion Waste and By-Products: MCR-ALS Applied to Reveal the Compositional Profiles of Alcoholic Fermentations of Onion Juice Monitored by Near-Infrared Spectroscopy. *Biotech Bioeng* 101: 776-787.

Grigelmo-Miguel, N. & Martin-Belloso, O. (1999). Comparison of Dietary Fiber from by-products of processing fruits and greens and from cereals. *Lebensm-Wiss u-Technol* 32: 503-508.

Hirota, S.; Shimoda, T.; Takahama, U. (1998). Tissue and Spatial Distribution of Flavonol and Peroxidase in Onion Bulbs and Stability of Flavonol Glucosides during Boiling of the Scales. *J Agric Food Chem* 46: 3497-3502.

Jaime, L.; Martín-Cabrejas, M.A.; Mollá, E.; López-Andréu, F.J.; Esteban, R.M. (2001a). Effect of storage on fructan and fructooligosaccharide of onion (Allium cepa L.*). J Agric Food Chem* 49: 982-988.

Jaime, L.; Martínez, F.; Martín-Cabrejas, M.A.; Mollá, E.; López-Andréu, F.J.; Waldron, K.W.; Esteban, R.M. (2001b). Study of total fructan and fructooligosaccharide content in different onion tissues. *J Sci Food Agric* 81: 177-182.

Jaime, L.; Mollá, E.; Fernández, A.; Martín-Cabrejas, M.A.; López-Andréu, F.J.; Esteban, R.M. (2002). Structural carbohydrate differences and potencial source of dietary fiber of onion (Allium cepa L.) tissues. *J Agric Food Chem* 50: 122-128.

Jung, Y.S.; Kim, M.H.; Lee, S.H.; Baik, E. J.; Park, S. W.; Moon, C.H. (2002). Antithrombotic effect of onion in streptozotocin-induced diabetic rat. *Prostaglandins, Leukotrienes Essent Fatty Acids* 66: 453–458.

Lanzotti, V. (2006). The analysis of onion and garlic. *J Chromatogr A* 1112: 3-22.

Le Marchand, L. (2002). Cancer preventive effects of flavonoids;a review. *Biomed Pharmacother* 56: 296–301.

Mallor, C. & Thomas, B. (2008). Resource allocation and the origin of flavour precursors in onion bulbs. *J Hortic Sci Biotech* 83(2): 191–198.

Matikkala, E.J. & Virtanen, A.I. (1967). On the quantitative determination of the amino acids and glutamyl peptide of onion. *Acta Chem Scand* 21: 2891-2893.

Olsson, M. E.; Gustavsson K.E.; Vagen, I.M. (2010). Quercetin and Isorhamnetin in Sweet and Red Cultivars of Onion (Allium cepa L.) at Harvest, after Field Curing, Heat Treatment, and Storage. *J Agric Food Chem* 58: 2323–2330.

Patil, B.S. & Pike, L.M. (1995). Distribution of quercetin content in different rings of various coloured onion (Allium cepa L.) cultivars. *J Hort Sci* 70: 643-650.

Prakash, D.; Singh, B.N.; Upadhyay, G. (2007). Antioxidant and free radical scavenging activities of phenols from onion (Allium cepa). *Food Chem* 102: 1389-1393.

Randle, W.M.; Lancaster, J.E.; Shaw, M.L.; Sutton, K.H.; Hay, R.L.; Bussard, M.L. (1995). Quantifying Onion Flavour Compounds Responding to Sulfur Fertility-Sulfur Increases Levels of Alk(en)yl Cysteine Sulfoxides and Biosynthetic Intermediates. *J Am Soc Hort Sci* 120(6): 1075-1081.

Randle, WM. Onion flavor chemistry and factors influencing flavor intensity. In: Risch SJ, Ho CT editors. Spices: Flavor Chemistry and Antioxidant Properties. ACS Symposium Series 660. Washington DC: *American Chemical Society;* 1997; 41– 52.

Rohn, S.; Buchner, N.; Driemel, G.; Rauser, M.; Kroh, L.W. (2007). Thermal Degradation of Onion Quercetin Glucosides under Roasting Conditions. *J Agric Food Chem* 55(4): 1568-1573.

Roldán, E.; Sánchez-Moreno, C.; de Ancos, B.; Cano, M.P. (2008). Characterisation of onion (Allium cepa L.) by-products as food ingredients with antioxidant and antibrowning properties. *Food Chem* 108: 907–916.

Roldán-Marín, E.; Sánchez-Moreno, C.; Lloría, R.; de Ancos, B.; Cano, M.P. (2009). Onion high-pressure processing: Flavonol content and antioxidant activity. *LWT-Food Sci Technol* 42: 835–841.

Rose, P.; Whiteman, M.; Moore, P.K.; Zhu, L.Z. (2005). Bioactive S-alk(en)yl cysteine sulfoxide metabolites in the genus Allium: the chemistry of potential therapeutic agents. *Nat Prod Rep* 22: 351– 368.

Santas, J.; Carbó, R.; Gordon, M.H.; Almajano, M.P. (2008). Comparison of the antioxidant activity of two Spanish onion varieties. *Food Chem* 107: 1210-1216.

Schieber, A.; Stintzing, F.C.; Carle, R. (2001). By-products of plant food processing as a source of functional compounds-recent developments. *Trends Food Sci Technol* 12: 401-413.

Slimestad, R. & Vågen I.M. (2009). Distribution of non-structural carbohydrates, sugars, flavonols and pyruvate in scales of onions, Allium cepa L. *J Food Agric Environ* 7: 289-294.

Thomas, D.J. & Parkin, K.L. (1994). Quantification of Alk(en)yl-L-cysteine Sulfoxides and Related Amino Acids in Alliums by High-Performance Liquid Chromatography. *J Agric Food Chem* 42: 1632-1638.

Trammell, K.W. & Peterson, C.E. (1976). Quantitative differences in the flavonol content of yellow onion, Allium cepa L. *J Amer Soc Hort Sci* 101: 205-207.

Tsushida, T. & Suzuki, M. (1995). Isolation of flavonoid-glycosides in onion and identification by chemical synthesis of the glycosides. *Nipp Shok Kaga Kaishi* 42: 100–108.

Waldron, K.W. (2001). Useful ingredients from onion waste. *Food Sci Techol* 15(2): 38-41.

Yamada, K.; Naemura, A.; Sawashita, N.; Noguchi, Y.; Yamamoto, J. (2004). An onion variety has natural antithrombotic effect as assessed by thrombosis/thrombolysis models in rodents. *Thrombosis Res* 114: 213–220.

Yang, J.; Meyers, K.J.,;Van Der Heide, J.; Liu, R.H. (2004). Varietal Differences in Phenolic Content and Antioxidant and Antiproliferative Activities of Onions. *J Agric Food Chem* 52 (22): 6787-6793.

Yao, L.H.; Jiang, Y.M.; Shi, J.; Tomás-Barberán, F.A.; Datta, N.; Singanusong, R.; Chen, S.S. (2004). Flavonoids in food and their health benefits. *Plant Foods Hum Nutr* 59: 113-122.

Yoo, K.S. & Pike, L.M. (1998). Determination of flavor precursor compound S-alk(en)yl-L-cysteine sulfoxides by an HPLC method and their distribution in Allium species. *Sci Hortic* 75: 1-10.

In: Onion Consumption and Health
Editors: C. B. Aguirre et al.
ISBN: 978-1-62100-836-1

Chapter 6

A COMPARISON TECHNIQUE FOR FLAVONOIDS EXTRACTION FROM ONIONS: ULTRASOUND ASSISTED EXTRACTION

Zill-e-Huma*[*], *Maryline Abert Vian and Farid Chemat
Université d'Avignon et des Pays de Vaucluse,
INRA, UMR Avignon, France

ABSTRACT

Due to the use of ambient temperature (25°C) and water having pH 7 during extraction step, ultrasounds have been found to be an interesting alternative to conventional methods for antoxidants extraction. Extracts obtained in presence of ultrasound waves have shown clearly their impact in increase of total dry extract yield along with increase extraction of flavonol contents. Another advantage of using ultrasounds is the rapidity of the extraction.

This extraction step require several hours in conventional processes, but it only required few minutes in ultrasounds assisted systems which induced better recovery of flavonol contents.

[*] E-mail: farid.chemat@univ-avignon.fr.

1. INTRODUCTION

1.1. Green Extraction Techniques

Shortcomings of existing extraction technologies, like increase consumption of energy (more than 70% of total process require energy), high rejection of CO_2 and more consumption of harmful chemicals, have forced the food and chemical industries to find new separation "green" techniques which typically use less solvent and energy, such as microwave extraction, supercritical fluid extraction, ultrasound extraction, ultrafiltration, flash distillation, the controlled pressure drop process and subcritical water extraction. Separation under extreme or non-classical conditions is currently a dynamically developing area in applied research and industry.

1.1.1. Microwave-Assisted Extraction (MAE)

Using microwaves, extraction and distillation can now be completed in minutes instead of hours with high reproducibility, reducing the consumption of solvent, simplifying manipulation and work-up, giving higher purity of the final product, eliminating posttreatment of waste water and consuming only a fraction of the energy normally needed for a conventional separation method such as distillation or solvent extraction. Several classes of compounds such as essential oils, aromas, pigments, antioxidants, and other organic compounds have been extracted efficiently from a variety of matrices (mainly animal tissues, food and plant materials). The advantages of using microwave energy, which is a non contact heat source, includes: more effective heating, faster energy transfer, reduced thermal gradients, selective heating, reduced equipment size, faster response to process heating control, faster start-up, increased production, and elimination of process steps [1-2].

1.1.2. Supercritical Fluid Extraction (SFE)

SFE is an innovative method with green theme behind its development for the extraction of solid materials using a supercritical fluid [5-6]. Carbon dioxide is the most widely used solvent in SFE because it is non-toxic, non-flammable, cheap, easily eliminated after extraction and endowed with a high solvating capacity for nonpolar molecules. Other possible solvents are Freon, ammonia and some organic solvents [7]. In a typical SFE procedure, the supercritical fluid continuously enters the solid matrix where it dissolves the material of interest. The extraction can be achieved with a remarkably high selectivity by adjusting the solvating capacity of the supercritical fluid by changing the pressure and/or temperature. Major advantages of SFE include preconcentration effects, cleanliness, safety and simplicity [8]. The drawbacks of SFE are the need of more expensive equipment with the difficulty of extracting polar molecules without adding modifiers to CO_2.

1.1.3. Pulsed Electric Field Extraction (PEF)

Application of pulsed electric field (PEF) either concurrently or before pressure application is an emerging extraction method, which has improved non thermal pressing efficiency by enhancing extraction of valuable components with adequate yield and purity of extracted juice in very short time with very less energy consumption [9]. PEF actually influence textural properties of plant material and improve tissue softness by electroporation of cell membrane resulted with rapid and enhance mass transfer. These short duration pulses of high intensity are not only applicable for extraction of juice at large scale but also permit extraction of sensitive coloring pigment at room temperature [10].

1.1.4. Instantaneous Controlled Pressure Drop (DIC, Détente Instanttanée Contrôlée)

"Instant Controlled Pressure-Drop" process (DIC: Détente Instantanée Contrôlée), is actually a swelling operation which was developed few years ago by Allaf and Louka [11]. The aim behind the development of this process was to improve the hydration capacity and dissolution rate of dry fruits and vegetables, which normally oppose resistance to penetration of solvent due to their natural structure. The matrix is subjected to an abrupt transition from high steam pressure to close to vacuum, resulted with plant product with more expended structure. Initially it was used for food industries especially for drying purpose but laterally it was also used for extraction of volatile components and essential oil from different plant parts including fruits and seeds. Recently the effect of this process was also studied for anthocyanin extraction from Roselle calyces which resulted with increase yield due to higher overall diffusivity [12].

1.1.5. Ultrasound-Assisted Extraction (UAE)

Ultrasound assisted extraction is an emerging potential technology that can accelerate heat and mass transfer and has been successively used in extraction field. Ultrasound waves after interaction with subjected plant material alter its physical and chemical properties and their cavitational effect facilitates the release of extractable compounds and enhances the mass transport by disrupting the plant cell walls. UAE is a clean method that avoids the use of large quantity of solvent along with cutting down in the working time. Ultrasounds are successively employed in plant extraction field [3-4].

1.2. Applications of Ultrasound in Food Technology

In pre-historic ages, the competition of food urged the processing of food to preserve it for longer times. Nowadays, the processed foods that are flourishing in supermarkets are modern processed foods and traditional foods, but their manufacturing, processing and packaging technologies have been advanced and rationalized to an incomparable extent. The principle aims of these technologies are to reduce the processing time, save energy and improve the shelf life and quality of food products. Thermal technologies (radio frequency and microwave heating), vacuum cooling technology, high pressure processing and pulsed electric field technology are those novel technologies who have potential for producing high-quality and safe food products but current limitations related with high investment costs, full control of variables associated with the process operation, lack of regulatory approval and importantly consumer acceptance have been delaying a wider implementation of these technologies at the industrial scale. In recent years, ultrasound (US) in the food industry has been the subject of research and development. There is a great interest in ultrasound due to the fact that industries can be provided with practical and reliable ultrasound equipment. Nowadays, its emergence as green novel technology has also attracted the attention to its role in the environment sustainability. Ultrasound applications are based on three different methods:

- Direct application to the product
- Coupling with the device
- Submergence in an ultrasonic bath

There are a large number of potential applications of high intensity ultrasound in food processing e.g filtration [13], defoaming [14], degassing [15], depolymerisation, emulsification, cleaning and cutting; others are in use but still under development e.g. drying, freezing and sterilisation and several are of great potential interest but await full realisation e.g. extraction, fermentation and meat tenderisation. A key aspect of the application of ultrasonic methods is that they are often used in conjunction with existing technologies to provide more efficient treatments e.g. thermosonication and drying.

1.3. Ultrasounds Theory

Ultrasounds are mechanic waves that necessitate an elastic medium to spread over. The difference between sound and ultrasounds is the frequency of the wave : sounds are at human earring frequencies (from16 Hz to 16-20 kHz) while ultrasounds have frequencies above human earring but below microwaves frequencies (from 20 kHz to 10 MHz).

As a sound wave passes through an elastic medium, it induces a longitudinal displacement of particles. In fact, the source of the sound wave acts as a piston on the surface of the medium [16]. This results in a succession of compression and rarefaction phases into the medium. When the piston is in its opened position it induces a compression into the medium and when the piston is in its contracted (pull) position it creates a rarefaction phase.

The series of compression and rarefaction regions generated as the sound wave passes through the medium creates an acoustic pressure in this medium. The acoustic pressure into the medium is positive during the compression cycle and is negative during the rarefaction cycle. These variations of pressure during compression and rarefaction phases induce molecular movements involving constriction and distension between groups of molecules. During the rarefaction phase, the distance between contiguous molecules (d) increases. Every medium has a critical molecular distance: below this critical value, the liquid remains intact, but above this distance, the liquid would break down and voids can be generated into the liquid. In the case of ultrasounds, if the rarefaction cycle is strong enough, the distance (d) between contiguous molecules can reach or even exceed the critical molecular distance of the liquid.

The voids created into the medium are the cavitation bubbles which are responsive of ultrasonic effect. In fact theses cavitation bubbles are able to grow during rarefaction phases and decrease in size during compression cycles. When the size of these bubbles reach a critical point they collapse during a compression cycle and release large amounts of energy. The temperature and the pressure at the moment of collapse have been estimated to be up to 5000 K and 2000 atmospheres in an ultrasonic bath at room temperature. This creates hotspots that are able to accelerate dramatically the chemical reactivity into the medium.

When these bubbles collapse onto the surface of a solid material, the high pressure and temperature released generate microjets directed towards the solid surface. These microjets are responsible for the degreasing effect of ultrasounds on metallic surfaces which is widely used for cleaning materials. Another application of microjets in food industry is the extraction of vegetal compound. As shown in figure 25, a cavitation bubble can be generated close to the plant material surface, then during a compression cycle, this bubble collapse and a microjet directed toward the plant matrix is created. The high pressure and temperature involved in this process will destroy the cell walls of the plant matrix and its content can be released into the medium. This is a very interesting tool for ingredient extraction from natural products.

1.4. Application of US in Extraction Field

The use of ultrasound can enhance the extraction process by increasing the mass transfer between the solvent and plant material. The collapse of cavitation bubbles leads to better cell disruption through the formation of microjets due to assymetrical bubble collapse near a solid surface. This allows for improved solvent penetration into the plant body itself and can also break down cell walls [17]. As a consequence, employing ultrasound in the use of plant extraction has benefits in increased mass transfer, better solvent penetration, less dependence on solvent used, extraction at lower temperatures, faster extraction rates and greater yields of product [18]. These features make sonication an attractive proposition for many extractions and for scale-up it should be used in the extraction unit itself where the plant material is in direct contact with the solvent. Flavour and fragrances are complex mixtures of volatile compounds usually present in low concentrations. They are present in varieties of aromatic plants, either in the roots, stems, seeds, leaves, flowers or fruits.

The molecules responsive for the flavour of the aromatic plant can be extracted and used in the food industry, medicine or perfumeries. The first aroma extractions are difficult to date precisely. Old testimonies allow concluding that the Hindus controlled fermentation and obtained odorant oils starting from rudimentary distilling apparatuses. Now a days, this distillation process has been improved by its coupling with ultrasound or even the development of new ultrasoundassisted extraction (UAE). Literature reported the extraction of different aroma components from aged brandies [19], tea [20] and wine [21] by ultrasound baths. UAE allowed better yields and a shorter extraction time compared to conventional methods.

UAE has also been developed for essential oil from aromatic plants such as peppermint leaves [22], artemisia [23] and lavender [24], or from other vegetal matrix such as garlic [25] and citrus flowers [26]. Increased yields of essential oil were found for peppermint leaves (up to 12%) and for artemisia when using UAE, and increase by 2 to 3 fold of the main compounds of lavandula essential oil when comparing UAE to conventional distillation.

Moreover, UAE not only improved yields but as the method is fast and run at low temperature, the final product usually showed less thermal degradation than traditional methods. Several studies have been run out on extraction of the main aroma compounds from spices. For example the vanillin was extracted from vanilla pods [27] carvone from caraway seeds [28] and safranal from Greek saffron [29]. Yields of vanillin were comparables after one hour by UAE vs 8 h in conventional extraction.

Another publication showed that 80% of the pure vanillin was obtained after only 120 sec of ultrasounds (ultrasonic probe) whilst it took 24 hours with the conventional method to obtain 100% [30]. For carvone extraction, the UAE method was compared to the Soxhlet. Unwanted fatty materials were extracted by Soxhlet extraction while the extract with UAE was of better quality and richer in carvone than in limonene Wide varieties of fruits and vegetables have been studied by UAE because antioxidants are present in different amounts in different varieties of plants and these antioxidants come from different families. One of the most common antioxidant is the lycopene extracted from tomatoes [31]. In this particular work, authors not only worked on UAE but also on coupling ultrasounds with microwaves which gives high lycopene extraction in only 6 min.

Herrera et al. [32] showed that similar amounts of phenolic compounds could be extracted from strawberries in 2 minutes while it took 20 h with conventional method and 3 h using supercritical fluid extraction. Another example is the extraction of anthocyanins from raspberries developed by Chen et al. [33]. A fast UAE of capsaicinoids from pepper as also been set up by Barbero et al. [34] who developed a reproducible ultrasonic extraction method using methanol as solvent. Citrus by-products are not only used for aromatic compounds contained in the essential oils, but phenolic compounds are also extractable from citrus peels [35] by using ulrtrasound. Other phenolic compounds have been extracted from coconut shell [36] and from wheat bran [37]. Furthermore, it has a number of applications for extraction of polyphenols from different plant matrix but the literature search did not yield any reference about earlier reports on the UAE of phenolic compounds from onions just by using water as a solvent. The objective of this work is to use ultrasound as a novel comparative extraction technique for onion flavonols extraction.

2. Materials and Methods

2.1. Raw Material

Yellow onion (*Allium cepa* L.) by-products which were rejected by local supermarket in Avignon province (South France) were purchased. These onion by-products were used completely with their outer dry layer for their following processing after cutting in small slices.

2.2. Chemicals

All solvents used for chromatographic purposes were HPLC grade. Methanol and Formic acid were from Merck (Darmstadt, Germany) and Acetonitrile was from Fisher Scientific Ltd. (Bishop Meadow Road, Loughborough, UK).

The HPLC grade flavonol standards quercetin-3,4′-diglucoside and quercetin-4′-glucoside (spiraeoside) were purchased from Extrasynthese (Lyon, France). Quercetin and quercetin-3-glucoside were purchased from Sigma Chemicals Chimie (Fallavier, France). DPPH (2,2-diphenyl-1-picrylhydrazyl) was from Sigma-Aldrich (Germany).

2.3. Moisture Content Determination

Moisture content determination of onion by-products (whole bulb of onion including outer dry layer) was carried out firstly by conventional Dean-Stark distillation apparatus according to the official method of the American Oil Chemical Society [38] and also by dehydration in an electric oven at 80°C. The average moisture content measured by both processes was 84.5 ± 0.5%.

2.4. Extraction procedures

Ultrasound-assisted extraction (UAE) was performed with a PEX 3 Sonifier (R.E.U.S., Contes, France) composed of an inox jug having 23 * 13.7 cm internal dimensions with a maximal capacity of 3 L, and a transducer, in the base of jug, operating at a frequency of 25 kHz with maximum input power of 250 W. The double layered mantle allowed us to control the temperature of the medium by cooling/heating systems. The detailed diagram of the apparatus has shown in the Figure 1. Conventional maceration, made for comparison, was carried out in the exactly same conditions without ultrasound. Extraction experiments were performed in triplicate.

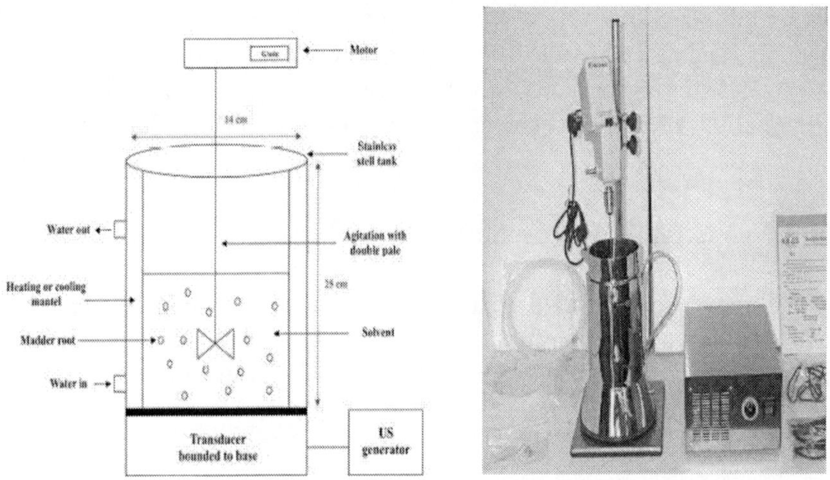

Figure 1. Laboratory sono-extraction reactor (3 liter system).

2.5. HPLC Analysis

HPLC analyses were performed using a Waters (Milford, MA) HPLC system consisting of a Waters 600E pump, a Waters 717 manual injector rheodyn, a Waters 2996 photodiode array detector. The HPLC pumps, manual injector rheodyn, column temperature, and diode array system were monitored and controlled by using Waters Empower 2 Chromatography Data software program. The wavelength used for the quantification of the onion flavonoids with the diode detector was 360 nm. The chromatographic separation was carried out on a Purospher Star RP-18 end-capped column (250 mm × 4 mm

I.D.; 5 μm particle size from VWR), with a RP18 guard column (4 mm×4mm I.D.; 5μm particle size also from VWR). The end-capped column and guard column were held at 37°C and the flow rate was set at 1mL/min. The mobile phase consisted of two solvents: (A) acidified water (0.5% formic acid) and (B) 100% acetonitrile. The solvent gradient used was the following: 0 min, (A) 95% and (B) 5%; 20min, (A) 60% and (B) 40%; 30min, (A) 0% and (B) 100%; 45min, (A) 95% and (B) 5%. The injection volume was 20 μL and all analyses were performed at least three times and only mean values were reported. Identification of flavonoids was done by comparing the elution order and UV-visible spectra. Quantification was carried out by using the external standards of known concentration. Peak areas were used to quantify the compounds in the sample. A linear regression analysis was carried out on the data of the peak area versus concentration. Linear calibration curves of the standards ranging from 10 to 100 mg/L were obtained with good linearity and R2 values which were more than 99.5% accurate for all the standards. Final concentrations of different flavonoids were calculated in mg/100g DW.

2.6. UV-Visible Spectroscopy

All absorbances of total phenolic contents (TPC) and DPPH assay were performed on a Hewlett-Packard 8453 diode array UV-vis spectrometer equipped with a magnetically stirred quartz cell (optical path length = 1 cm). The temperature in the cell was controlled by means of a water thermostated bath [39].

2.6.1. Determination of the Total Content of Reducing Compounds (TCRC)

TPC was estimated colorimetrically using the Folin-Ciocalteu method [40], with a kit (SEPPAL (Isitec-lab), France) especially suitable for TPC measurement of food products. This kit includes reagent A (modified Folin-Ciocalteu reagent), reagent B (alkaline buffer) and a gallic acid solution (3 g/L). A small volume (20 μL) of H2O (blank), gallic acid solution (standard) and 200 μL the extract (sample) was mixed with reagent A (2 mL). After 1 min, 1 mL of reagent B was added in both water and gallic acid standard and 850 μL in sample. The mixtures were allowed to stand for 30 min in the dark at room temperature. Then, their absorbance was measured at 760 nm. TPC were calculated by using the following formula:

TPC = 3 × (sample absorbance – blank absorbance) / (standard absorbance – blank absorbance). TPC measurements were performed thrice and mean values, expressed as mg gallic acid/g of dry weight (mg GAE/g DW), were reported.

2.6.2. DPPH Assay

DPPH is a stable highly coloured free radical that can abstract labile hydrogen atoms from phenolic antioxidants with simultaneous formation of a colourless hydrazine (DPPH-H) [41]. The radical scavenging activity of extracts was evaluated according to the method described by Monzocco [42], with some modifications. About 200 μL of a solution of onion extract in water were added to 2 mL of a DPPH solution (6x10-5 M) in MeOH and the mixture was left in the dark at room temperature for 30 min. The decrease in absorbance was measured at 517 nm. The percentage of DPPH consumed was calculated from the following equation: I % = 100 x (A0* A1) / A0, where A0 was the initial absorbance (no antioxidant) and A1 the absorbance in the presence of the extract or standard at different concentrations. The IC50 of each extract or standard (concentration giving I % = 50) was calculated by plotting I % as a function of the extract concentrations. All experiments were performed in triplicate.

3. Results and Discussion

3.1. Preliminary Study

Various experiments aim to optimize and evaluate phenolic compounds extraction in a convenient time, with appropriate solid/liquid was carried out in the preliminary study without ultrasonic assistance. Solvent we have used for extraction was just water at pH 7 at ambient temperature 25°C with theme of solvent free cold extraction.

3.1.1. Solid/Liquid Ratio

The solid/liquid ratio was optimized, to get maximum amounts of polyphenols, 5 to 30g of onion slices were extracted by conventional maceration method in 100mL of water. TPC concentrations were found to increase with the solid/liquid ratio (Figure 2) upto 25g/100mL, further increase of solid contents doesn't affect on more recovery of phenolic contents.

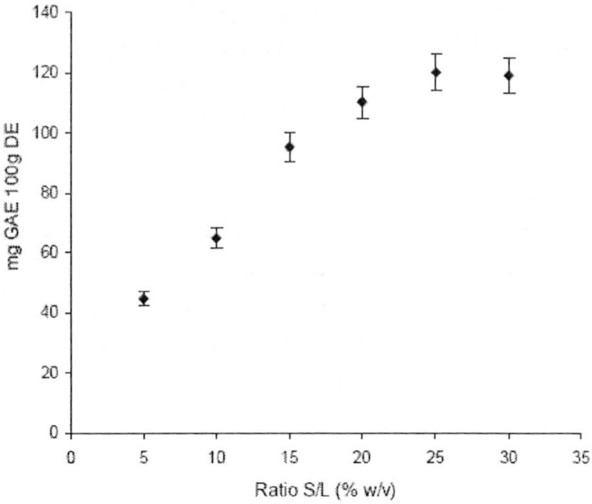

Figure 2. Effect of different solid liquid ratios on phenolic contents.

3.1.2. Power Optimization

After selection of solid/liquid ratio we have performed experiments at maximum ultrasound power 250W to determine the time limit to get the maximum phenolic contents. We have got maximum phenolic contents in 30mins of extraction, with further passage of time there was no remarkable increase of phenolic contents was detected. To optimize the power for extraction of onion flavonoids, we have analyzed various powers including 30W, 60W, 90W, 150 and 250W for total dry extract yield and phenolic contents and we have got maximum recovery of yield 1.95% at power of 250W. This obtained yield was clearly more than we have got with conventional maceration method (0.98%) at same conditions but in absence of ultrasound waves.

3.2. Comparison of US vs Conventional Method

Moving towards green extraction, we have tried to minimise the loss of nutritionally important components from onion by-products by extracting them with ultrasound waves at ambient temperature (25°C) by using water having pH 7. We have analysed the extracts obtained by UAE and conv. method for quantification of flavonols by HPLC, the data revealed the presence of quercetin-3,4′ -diglucoside and quercetin 4′ -glucoside as predominant form along with possessing higher concentration of quercetin aglycone (Figure 3) in both of extracts (Table 1).

The application of ultrasound waves resulted with increased yields of these flavonols from 401.3 to 449.29 for quercetin-3,4′ -diglucoside , from 163.2 to 230.5 for quercetin 4′ - glucoside and from 22.40 mg to 71.52 mg per 100 g dry extract for quercetin aglycone. Higher concentration of these flavonols in UAE extracts in comparison with their content obtained by traditional extraction process at same conditions reveals clearly the effectiveness of this process in the field of antioxidants extraction.

Table 1. Flavonol contents obtained by UAE and conventional maceration extraction process

Extraction method	QA* (mg/100gDW) 23.81 min	QDG* (mg/100gDW) 16.18 min	Q4G* (mg/100gDW) 19.85 min	Q3G* (mg/100gDW) 17.88 min	Total quercetin (mg QE/100g DW)
UAE	71.52 ± 3.8	449.29 ± 17.1	230.5 ± 7.8	18.9 ± 0.59	450.5
Conv	22.40 ± 1.1	401.3 ± 24.5	163.2 ± 5.6	13.8 ± 0.83	345.6

To found the specific impact of ultrasound, extracts (ultrasound and conventional) (figure 4) were compared for their total phenolic contents and antioxidant activity by DPPH assay.

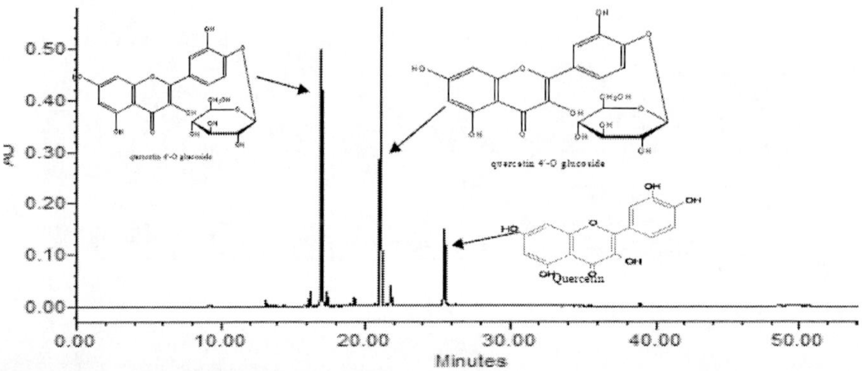

Figure 3. HPLC Chromatogram of ultrasound extracted onion byproducts.

Ultrasound-assisted extraction lead to a yield of gallic acid equivalents (TPC) (121 ± 3.8 mg GAE/g DW) increased by more than 20% than the yield obtained by conventional maceration method (89.6 ± 2.3mg GAE/g DW) in 30mins of extraction. The kinetic of extraction was clearly improved, which could be attributed to ultrasonic cavitation while it is the only variable added between both experiments. Chemat et al. showed, by-using electronic microscopy, physical effects of ultrasound on cells when applied on plant material [43].

Total phenolics generally correlate with antioxidant capacities displayed by the DPPH assay. Radical scavenging activities of both extracts in terms of their IC_{50} values are calculated. The DPPH method is one of the oldest and most frequently used methods as it provides a convenient and rapid way to evaluate antiradical activities of antioxidants. For measuring the antioxidant activity we have chosen the concentration of samples providing 50% inhibition (IC_{50}) of DPPH solution. Both extracts have given linear response above then R_2=0.9900 in the concentration range 1-200mg/ml. Similar to the total phenolic content, ultrasound extracted sample was more effective with IC_{50} value 10.01mg/ml against IC_{50} value of conventional extract (13.21mg/ml). Among newer techniques used in extraction technology, ultrasound assisted extraction (UAE) of food components has been employed as a new tool to improve the yield and quality of extraction products and to reduce the duration of procedures.

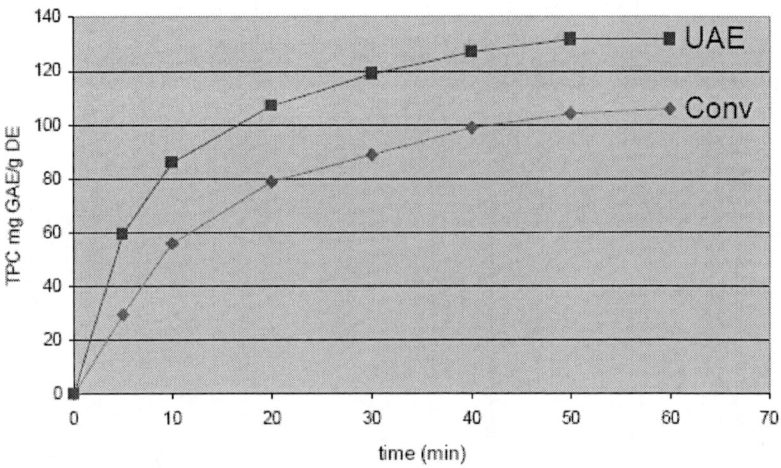

Figure 4. Comparison of UAE vs CSE on phenolic contents extraction.

Table 2. Advantages and drawbacks of traditional and recent extraction techniques

	MHG	UAE	MAE	SFE	ASE
Name	Microwave hydrodiffusion and gravity	Ultrasound-Assisted Extraction	Microwave Assisted Extraction	Supercritical Fluid Extraction	Accelerated Solvent Extraction
Brief description	Sample is directly put in the reactor and subjected to microwave irradiations	Sample is immersed in solvent and submitted to ultrasound using a US probe or US bath.	Sample is immersed in solvent and submitted to microwave energy.	Sample is placed in a high pressure vessel and crossed continuously by the supercritical fluid.	Sample is heated by a conventional oven and crossed by the extraction solvent under pressure.
Extraction time	15-30 min	10-60 min	3-30 min	10-60 min	10-20 min
Sample size	150g-1kg	1-30 g	1-10 g	1-5 g	1-30 g
Solvent use	No solvent	50-200 ml	10-40 ml	2-5 ml (solid trap)30-60 ml (liquid trap)	15-60 ml
Investment	Moderate	Low	Moderate	High	High
Advantages	Rapid Easy to handle No filtration required No solvent consumption	Easy to use	Rapid Easy to handle Moderate solvent consumption	Rapid Low solvent consumption Concentration of the extract No filtration necessary Possible high selectivity	Rapid No filtration necessary Low solvent consumption
Drawbacks	Appropriate selection of extraction parameters is its critical step, less dry extract yield	Large amount of solvent consumption Filtration step required	Extraction solvent must absorb microwave energy Filtration step required	Many parameters to optimize	Possible degradation of thermolabile analytes

UAE is a clean method that avoids the use of large quantity of solvent; this reduced environmental impact of UAE is clearly advantageous in terms of energy and time. UAE makes use of physical and chemical phenomena that are fundamentally different from those applied in conventional extractiontechniques. The present study reports on the extraction of polyphenols especially flavonols from onion (*Allium cepa* L.) by-products range by using water having pH 7 as solvent. The high total phenolic content and extraction yield obtained from optimised UAE proved its efficiency when compared with the conventional method. Furthermore, the antioxidant activity determined by the DPPH test confirmed the suitability of UAE for the preparation of antioxidant-rich plant extracts.

Finally we have tried to make the comparison by comparing advantages and disadvantages of MHG with other recent and traditional extraction methods like UAE, Supercritical Fluid Extraction (SFE), Accelerated Solvent Extraction (ASE) and Microwave- Assisted Extraction (MAE) in Table 2.

CONCLUSION

Due to the use of ambient temperature (25°C) and water having pH 7 during extraction step, ultrasounds have been found to be an interesting alternative to conventional methods for antoxidants extraction. Extracts obtained in presence of ultrasound waves have shown clearly their impact in increase of total dry extract yield along with increase extraction of flavonol contents. Another advantage of using ultrasounds is the rapidity of the extraction. This extraction step require several hours in conventional processes, but it only required few minutes in ultrasounds assisted systems which induced better recovery of flavonol contents.

REFERENCES

[1] Chemat, F; Lucchesi, M. Microwave-assisted extraction of essential oils, in Microwaves in organic synthesis, A. Loupy (ed) Weinheim: WILEY-VCH GmbH and Co. KGaA; 2006, 959-983.

[2] Pare, J; Belanger, J. Microwaves-Assisted Process (MAP): *principles and applications, in Instrumental methods in food analysis,* J. Pare, J. Belanger (eds.), Elsevier Sciences BV, Amsterdam, 1997, 395-420.

[3] Luque-García, J.L; Luque De Castro, M.D. Ultrasound: A powerful tool for leaching. *Trends in Analytical Chemistry*, 2003, 41 - 47.

[4] Lopez-Avila, V; Young, R; Teplitsky, N. Microwave-assisted extraction as an alternative to Soxhlet, Sonication, and Supercritical Fluid Extraction, *Journal of AOAC International*, 1996, 79, 142 – 156.

[5] Pourmortazavi, SM; Hajimirsadeghi, SS. Supercritical fluid extraction in plant essential and volatile oil analysis. *Journal of Chromatography A*, 2007, 1163, 2–24. [6] Sahena, F; Zaidul, ISM; Jinap, S; Karim, AA; Abbas, KA; Norulaini, NAN; Omar, AKM. Application of supercritical CO2 in lipid extraction – A review. *Journal of Food Engineering*, 2009, 95, 240–253.

[6] Luque de Castro, MD; Valcárcel, M; Tena, MT. *Analytical Supercritical Fluid Extraction*, Springer-Verlag, New York, 1994.

[7] Luque de Castro, MD; Jiménez-Carmona, MM. Where is supercritical fluid extraction going? *Trends in Analytical Chemistry,* 2000, 19, 223 - 228.

[8] Jemai, AB; Vorobiev, E. Pulsed Electric Field Assisted Pressing of Sugar Beet Slices: towards a Novel Process of Cold Juice Extraction. *Biosystems Engineering,* 2006, 93 (1), 57–68.

[9] López, N; Puértolas, E; Condón, S; Raso, J; Alvarez, I. Enhancement of the extraction of betanine from beetroot by pulsed electric fields. *Journal of Food Engineering*, 2009, 90, 60–66.

[10] [11] Karim, A; Nicolas, L; Francis, P; Jean-marie, B; Michel, F. Method for processing phytogenic materials to change their texture, apparatus therefore and resulting material. PCT/FR1994/000975, 1995.

[11] Amor, BB; Allaf, K. Impact of texturing using instant pressure drop treatment prior to solvent extraction of anthocyanins from Malaysian Roselle (Hibiscus sabdariffa). *Food Chemistry*, 2009, 115, 820–825.

[12] Smythe, MC; Wakeman, RJ. "The use of acoustic fields as a filtration and dewatering aid." *Ultrasonics,* 2000, 38(1-8), 657-661.

[13] Gallego Juarez, JA. Some applications of air-borne power ultrasound to food processing. Ultrasound in food processing. M. J. W. Povey and T. J. Mason. Glasgow, UK, Blackie Academic and Professional: 1998, 127–143.

[14] Mettin, R. Bubble structures in acoustic cavitation. Bubble and Particle Dynamics in Acoustic Fields: Modern Trends and Applications. A. A. Doinikov. Kerala (India), Research Signpost: 2005, 1–36.

[15] Mason, TJ. Chemistry with ultrasound, *Elsevier Applied Science,* New York, 1990.

[16] Toma, M; Vinatoru, M; Paniwnyk, L; Mason, TJ. Investigation of the effects of ultrasound on vegetal tissues during solvent extraction. *Ultrasonics Sonochemistry*, 2001, 8(2), 137-142.

[17] Vinatoru, M. An overview of the ultrasonically assisted extraction of bioactive principles from herbs. *Ultrasonics Sonochemistry,* 2001, 8(3), 303-313.

[18] Caldeira, I; Pereira, R; Climaco, MC; Belchior, AP; Bruno de Sousa, R. Improved method for extraction of aroma compounds in aged brandies and aqueous alcoholic wood extracts using ultrasounds. *Analitica Chimica Acta*, 2004, 125 – 134.

[19] Xia, T; Shi, S; Wan, X. Impact of ultrasonic-assisted extraction on the chemical and sensory quality of tea infusion. *Journal of Food Engineering*, 2006, 74, 557 – 560.

[20] Cabredo-Pinillos, S; Cdron-Fernandez, T; Gonzalez-Briongos, M; Puente-Pascual, L; Saenz-Barrio, C. Ultrasound-assisted extraction of volatile compounds from wine samples: Optimisation of the method. *Talanta*, 2006, 69, 1123 – 1129.

[21] Shotipruk, A; Kaufman, PB; Wang, HY; Feasability study of repeated harvesting of menthol from biologically viable Mentha x piperata using ultrasonic extraction. *Biotechnology progress*, 2001, 17, 924 – 928.

[22] Asfaw, N; Licence, P; Novitskii, AA; Poliakoff, M. Green chemistry in Ethiopia: the cleaner extraction of essential oils from Artemisia afra: a comparison of clean technology with conventional methodology. *Green Chemistry*, 2005, 7, 352 – 356.

[23] Da Porto, C; Decorti, D; Kikic, I. Flavour compounds of Lavandula angustifolia L., to use in food manufacturing: Comparison of three different extraction methods. *Food Chemistry*, 2009, 112, 1072 – 1078.

[24] Kimbaris, AC; Siatis, NG; Daferera, DJ; Tarantilis, PA; Pappas, CS; Polissiou, MG. Comparison of distillation and ultrasound-assisted extraction methods for the isolation of sensitive aroma compounds from garlic (*Allium sativum*). *Ultrasonics sonochemistry*, 2006, 13, 54 – 60.

[25] Alissandrakis, E; Daferera, D; Tarantilis, PA; Polissiou, M; Harizanis, PC. Ultrasoundassisted extraction of volatile compounds from citrus flowers and citrus honey. *Food Chemistry*, 2003, 82, 575 – 582.

[26] Jadhav, D; Rekha, BN; Gogate, PR; Rathod, VK. Extraction of vanillin from vanilla pods: A comparison study of conventional soxhlet and ultrasound-assisted extraction. *Journal of Food Engineering*, 2009, doi:10.1016/j.jfoodeng.2009.02.007.

[27] Chemat, S; Lagha, A; AitAmar, H; Bartels, PV; Chemat, F. Comparison of conventional and ultrasound-assisted extraction of carvone and limonene from caraway seeds. *Flavour and fragrance journal*, 2004, 19, 188 – 195.

[28] Kanakis, CD; Daferera, DJ; Tarantilis, PA; Polissiou, MG. Qualitative determintaion of volatile compounds and quantitative evaluation of safranal and 4-Hydroxy-2,6,6,-trimethyl123 1-cyclohexane-1-carboxaldehyde (HTCC) in Greek Saffron. *Journal of Agriculture and Food Chemistry*, 2004, 52, 4515 – 4521 .

[29] Sharma, A; Verma, SC; Saxena, N; Chadda, N; Pratap Singh, N; Kumar Sinha, A. Microwave and ultrasound-assisted extraction of vanillin and its quantification by high performance liquid chromatography in Vanilla planifolia. *Journal of Separation Science*, 2006, 29, 613 – 619.

[30] Liangfu, Z; Zelong, L. Optimization and comparison of ultrasound/microwave assisted extraction (UMAE) and ultrasonic assisted extraction (UAE) of lycopene from tomatoes. *Ultrasonics sonochemistry*, 2008, 15, 731 – 737.

[31] Herrera, MC; Luque de Castro, MD; Ultrasound-assited extraction of phenolic compounds from strawberries prior to liquid chromatographic separation and photodiode array ultraviolet detection. *Journal of Chromatography A*, 2005, 1100, 1–7.

[32] Chen, F; Sun, Y; Zhao, G; Liao, X; Hu, X; Wu, J; Wang, Z. Optimization of ultrasoundassited extraction of anthocyanins in red raspberries and identification of anthocyanins in extract using high-performance liquid chromatography-mass spectrometry. *Ultrasonics sonochemistry*, 2007, 14, 767– 778.

[33] Barbero, GF; Liazid, A; Palma, M; Barroso, CG. Ultrasound-assisted extraction of capsaicinoids from peppers. *Talanta*, 2008, 75, 1332 – 1337.

[34] Ma, YQ; Chen JC; Liu, DH; Ye, XQ. Simultaneous extraction of phenolic compounds of citrus peel extracts: Effect of ultrasound. *Ultrasonics sonochemistry*, 2009, 16, 57 – 62.

[35] Rodrigues, S; Pinto, GAS; Fernandes, FAN. Optimization of altrasound extraction of phenolic compounds from coconut (*cocos nucifera*) shell powder by response surface methodology. *Ultrasonics sonochemistry*, 2008, 15, 95 – 100.

[36] Wang, J; Sun, B; Cao, Y; Tian, Y; Li, X. Optimisation of ultrasounds-assisted extraction of phenolic compounds from wheat bran. *Food chemistry*, 2008, 106, 804 – 810.

[37] Official Method Ja 2a–46, American Oil Chemist Society, *Champaign,* IL, 1993.

[38] Dufour, C; Loonis, M; Dangles, O. Inhibition of the peroxidation of linoleic acid by the flavonoid quercetin with in their complex with human serum albumin. *Free Radical Biology and Medicine.* 2007, 43, 241–252.

[39] Vinson, JA; Hontz, BA. Phenol antioxidant index: Comparative antioxidant effectiveness of red and white wines. *Journal of Agriculture and Food Chemistry,* 1995, 43, 401–403.

[40] Diouf, PN; Stevanovic, T; Cloutier, A. Study on chemical composition, antioxidant and anti-inflammatory activities of hot water extract from Picea mariana bark and its proanthocyanidin-rich fractions. *Food Chemistry,* 2009, 113, 897–902.

[41] Monzocco, L; Anese, M; Nicoli, MC. Antioxidant properties of tea extracts as affected by processing. *Lebensm. Wiss. Technol.* 1998, 31, 694–698.

[42] Chemat, S; Lagha, A; AitAmar, H; Bartels, PV; Chemat, F. Comparison of conventional and ultrasound-assisted extraction of carvone and limonene from caraway seeds. *Flavour Fragrance Journal*, 2004, 188–195.

INDEX

B

C

D

Q

R

S